CONDI vs. HILLARY

ALSO BY DICK MORRIS

Because He Could
(with Eileen McGann)

Rewriting History
(with Eileen McGann)

Off with Their Heads

Power Plays

Vote.com

The New Prince

Behind the Oval Office

Bum Rap on American Cities

CONDI VS. HILLARY

THE NEXT GREAT PRESIDENTIAL RACE

DICK MORRIS | AND EILEEN MCGANN

ReganBooks
An Imprint of HarperCollins*Publishers*

HarperCollins books may be purchased for educational, business, or sales promotional use. For information please write: Special Markets Department, HarperCollins Publishers Inc., 10 East 53rd Street, New York, NY 10022.

Designed by Publications Development Company of Texas

ISBN: 0-06-083913-9

In memory of Genevieve "Jeanne" McGann

1927–2005

CONTENTS

ACKNOWLEDGMENTS

Thanks to our agent, Joni Evans, for her constant encouragement and support.

We are grateful to Katie Vecchio for her help in research, editing, and proofing. She is a gem—and a disciplinarian!

Tom Gallagher deserves an enormous thanks for his hard work in making sense of the endnotes.

And thanks to Maureen Maxwell, for tirelessly scheduling hundreds of interviews throughout the years for all of our books.

Calvert Morgan, after editing five of our books, can take credit not just for improving the book but also for improving our writing style—and always with wit and grace.

Cassie Jones, Elizabeth Yarborough, Vivian Gomez, Paul Crichton, Melissa McCarthy, and Larry Pekarek at ReganBooks also deserve thanks.

And, finally, there is Judith Regan, our publisher. We especially appreciate her innate instinct for discerning what works and what doesn't.

1

Setting the Stage

"I, Hillary Rodham Clinton, do solemnly swear that I will faithfully execute the office of president of the United States and will, to the best of my ability, preserve, protect, and defend the Constitution of the United States, so help me God."

On January 20, 2009, at precisely noon, the world will witness the inauguration of the forty-fourth president of the United States. As the chief justice administers the oath of office on the flag-draped podium in front of the U.S. Capitol, the first woman president, Hillary Rodham Clinton, will be sworn into office. By her side, smiling broadly and holding the family Bible, will be her chief strategist, husband, and copresident, William Jefferson Clinton.

If the thought of another Clinton presidency excites you, then the future indeed looks bright. Because, as of this moment, there is no doubt that Hillary Clinton is on a virtually uncontested trajectory to win the Democratic nomination and, very likely, the 2008 presidential election. She has no serious opposition in her party. More important, a majority of *all* American voters—52 percent—now supports her candidacy.[1]

The order of presidential succession from 1992 through 2008, in other words, may well become Bush, Clinton, Bush, Clinton.

But if the very thought of four—or perhaps even eight—more years of the Clintons and their predictable liberal policies alarms you; if you see through the new Hillary brand—that easygoing, smiling moderate; if you remember what a partisan, ethically challenged, left-wing ideologue she has always been, is now, and will always be, then you can see what the future holds.

That's exactly the kind of president Hillary Clinton would be.

But her victory is not inevitable. There is one, and only one, figure in America who can stop Hillary Clinton: Secretary of State Condoleezza "Condi" Rice. Among all of the possible Republican candidates for president, Condi alone could win the nomination, defeat Hillary, and derail a third Clinton administration.

Condoleezza Rice, in fact, poses a mortal threat to Hillary's success. With her broad-based appeal to voters outside the traditional Republican base, Condi has the potential to cause enough major defections from the Democratic Party to create serious erosion among Hillary's core voters. She attracts the same female, African American, and Hispanic voters who embrace Hillary, while still maintaining the support of conventional Republicans.

This is a race Condi can win.

And Hillary cannot offset these losses of reliable Democratic constituencies with other voting blocs. White men don't like her. That won't change. And there is nowhere else for her to pick up support. It's simple: With Condi in the race, Hillary can't win.

The stakes are high. In 2008, no ordinary white male Republican candidate will do. Forget Bill Frist, George Allen, and George Pataki. Hillary would easily beat any of them. Rudy Giuliani and John McCain? Either of them could probably win, but neither will ever be nominated by the Republican Party. These two are too liberal, too maverick, to win the party's support; their positions are too threatening to attract the Republican base. Jeb Bush? Too many Bushes in a row make a hedge. He's not going anywhere. And

Austrian-born Arnold Schwarzenegger can't run. In the next election, none of the usual suspects can stop Hillary. Without Condi as her opponent, Hillary Clinton will effortlessly lead the Democratic Party back into the White House in 2008.

There is, perhaps, an inevitability to the clash: Two highly accomplished women, partisans of opposite parties, media superstars, and quintessentially twenty-first-century female leaders, have risen to the top of American politics. Each is an icon to her supporters and admirers. Two groundbreakers, two pioneers. Indeed, two of the most powerful women on the planet: *Forbes* magazine recently ranked Condi as number one and Hillary as number twenty-six in its 2005 list of the most powerful women in the world.

As Hillary and Condi emerge as their party's charismatic heroines, they seem fated to meet on the grand stage of presidential politics. These two forces, two vectors, two women, and two careers may be destined to collide on the ultimate field of political battle. Two firsts in history. But only one will become president.

The year 2008 could, at last, be the year of the woman—indeed, the year of two women. Suddenly, the timing is right. Eighty-five years after the Nineteenth Amendment gave women the right to vote, the planets seem suddenly aligned to challenge history. American voters are surprisingly ready for a woman in the White House. Public opinion is rapidly settling into a consensus that a woman could actually be elected president in the next election. For the first time in our history, a majority of voters say they would support a woman for president. In a May 2005 *USA Today*/CNN/Gallup Poll, an amazing 70 percent of the respondents indicated that they "would be likely to vote[2] for an unspecified woman for president in 2008."[3]

What a revolutionary shift in thinking! No major American political party has ever nominated a woman for president. And only one woman has run for vice president—Democratic Party nominee Geraldine Ferraro in 1984. But now there are two star-crossed, qualified, and visible women who may be presidential contenders in 2008. And the voters like them both:[4] 53 percent of those questioned in the May 2005 survey had a favorable opinion of Hillary Clinton, while 42 percent rated her negatively. Condoleezza Rice fared much better: 59 percent liked her and only 27 percent didn't.

Hillary Clinton has always wanted to be the first woman president of the United States. Shortly after her husband's election in 1992, the couple's closest advisers openly discussed plans for her eventual succession after Bill's second term. Of course, things didn't turn out quite that way; Hillary has had to wait a bit. But her election to the Senate in 2000 gave her the national platform she needed to launch her new image—the "Hillary Brand,"[5] as we called it in *Rewriting History*—and begin her long march back to the White House.

One thing is certain: Hillary Clinton does not want any other woman to take what she regards as her just place in history as America's first woman president.

Yet, ironically, it is Hillary's own candidacy that makes Condi's necessary and therefore likely. The first woman nominated by the Democrats can only be defeated by the first woman nominated by the Republicans. Two firsts of their kind, locked in electoral combat, with the future—theirs and ours—on the line.

Their potential battle recalls a moment in the Civil War when the South was suffocating beneath the blanketing blockade the Union had draped over its ports.[6] Anxious to redress the balance, Confederate shipbuilders fitted the captured wooden-hulled Union warship *Merrimack* with a coating of iron skin and renamed it the *Virginia*. On March 8, 1862, a new kind of vessel—the world's first ironclad ship—sailed out to sea. There, in a single day, it demonstrated graphically its manifest superiority over every other ship in history. The *Virginia* rammed and sank the huge Union ship *Cumberland* and shelled the frigate *Congress* until she surrendered.

Once there was a *Merrimack*, however, there also had to be a *Monitor*, the Union's answer to this strange new creature of the sea. And so the Northern counterpart materialized, looking like a tin can set on a raft. The two ships met in mortal combat—two firsts of their kind, each made necessary by the other's potential to master the high seas.

The battle between the world's first ironclads ended in a stalemate. A race between Hillary Clinton and Condi Rice will have a more decisive ending. But the parallel is clear: If there is a Hillary, there must be a Condi. One will spawn the other.

Hillary's nomination as the first woman candidate for president by a major political party would generate extraordinary excitement and give the Democrats an undeniable advantage in the general election. The Republicans would have no choice but to respond by nominating a similarly compelling and popular candidate—one who would counteract the certain shift of women voters to Hillary. And who else could that be but Condi?

Consider this: If Hillary is nominated as the first woman ever to run for president, she is very, very likely to win. By maximizing her support among the 54 percent of the vote that is cast by women—and tapping into the enthusiasm that her husband elicits among African Americans and Hispanics—she is likely to sweep into office, easily defeating any conventional white male candidate the Republicans might send against her.

And there is only one viable Republican answer to Hillary's candidacy: Condoleezza Rice.

Were Condi and Hillary to face one another, it would be the next great American presidential race and one of the classic bouts in history. Hector vs. Achilles. Wellington vs. Bonaparte. Lee vs. Grant. Mary, Queen of Scots vs. Elizabeth. Ali vs. Frasier. And now, Condi vs. Hillary.

But these potential combatants are as different as, well, black and white. In many ways, they are mirror images of each other: not only white/black but north/south; Democrat/Republican; married/single; suburban/urban; and, in their policy interests, domestic/foreign.

Their backgrounds are not in the least similar. While Hillary grew up in the middle-class security of white, Protestant Park Ridge, Illinois, Condi came of age on the wrong side of the racial divide in pre–civil rights Birmingham, Alabama. But growing up as an African American in the segregated South did not mean that Condi came from an impoverished background. It was Rice who came from an educated, professional family; Hillary's was far more blue-collar. Hillary's mother, the child of a teen pregnancy who was abandoned by her mother and raised by her grandmother, was a high school graduate; her father, a physical education major and football player at Penn State, made and sold commercial draperies. Condi's parents and grandparents, on the other hand, were college graduates. Her father was a minister, teacher, and guidance counselor. Her mother was also a professional, a music teacher in the same school where her husband taught.

It is not only their family backgrounds and geography that were distinctive. Their careers also took very different paths. For more than thirty years, Hillary's success has always been coupled with her relationship with one powerful man: Bill Clinton. Wherever he went, Hillary followed, supporting him, advising him, rescuing him, and, at the same time, reaping enormous rewards from his advancement. Her own talents were often obscured, her ambitions put aside, as the two worked jointly to advance his career above all else.

It was Bill who introduced her to his colleagues at the University of Arkansas Law School when she was suddenly unemployed after her work as a legal researcher on the Watergate Committee came to an end in 1973. Though a bright and talented graduate of Yale Law School, Hillary had failed the D.C. bar exam and would undoubtedly have had a hard time landing a top position in Washington. Women lawyers were not yet in strong demand, and a bar failure would have been a major strike against her, as well as a humiliating admission to make in job interviews for a supremely self-confident person like Hillary. An easy alternative was Arkansas, where she had passed the bar the previous year and had since been admitted to practice law. Her decision to move to Fayetteville, Arkansas, and accept a teaching position in a clinic handling criminal law—a subject in which she had never before shown any interest—changed her destiny and paired her future with Bill Clinton's. From then on, as Bill moved up in Arkansas politics, Hillary simultaneously progressed in her legal career. When he was elected attorney general, she was offered a job at the Rose Law Firm, the most prestigious in Arkansas. When he was elected governor, she was named the firm's first woman partner. And when he was elected president, she ultimately evolved into a Senate candidate from New York.

Unlike Hillary, Condi has never married, and her success has never been a matter of hitching her wagon to the political fortunes of any powerful man. Instead, she advanced strictly on her own merits. She began her career by excelling as an academic and specializing in foreign affairs. Eventually, she brought that expertise to a family of presidents. But it was always Condi's own record of accomplishment that made her a prominent national figure. When she was still in her twenties, she was elevated to the Stanford University faculty because she amazed her colleagues with her abilities. She came to Washington during the administration of President George H. W. Bush because she had impressed National Security Advisor Brent Scowcroft, who met her at Stanford. She was only thirty-four when she became the administration's chief expert on the Soviet Union. After her White House experience, she so impressed the incoming president of Stanford that he asked her to be his provost, even though the job usually went to a dean or a department chair. Through Ronald Reagan's secretary of state, George Schultz, she met then-governor George W. Bush, and prepared him for the foreign policy issues he would face in the 2000 campaign. The younger Bush was so awed by Condi's abilities that he appointed her national security advisor and then secretary of state.

Condi Rice, in short, reached her position of power on the strength of her own achievements.

The two women also came to the White House in characteristically different ways. Hillary arrived as a wife, with no experience in government, no portfolio, no administrative experience. Though her husband immediately granted her sweeping authority over health care, she was still the president's wife, the first lady, who had no expertise in the very health care issues that she completely controlled. Her power was always derivative. She was not an elected official. She was not a cabinet member. She had no designated role or powers. The public policy issues she chose to address were centered on traditional women's issues: health care, advocacy for women and children, and protection of national treasures.

Rice entered the White House in a completely different way. She came in as a high-level expert, charged with guiding America through the delicate process of German reunification, the dismantling of the Soviets' satellite empire in Eastern Europe, and the eventual breakup of the Soviet Union itself. A rare woman in a field long dominated by men, she held her own.

The work these two women did once in the White House likewise reflected their dramatically opposite characters. Condi quietly advanced and enhanced her reputation in the field of national security and Soviet relations with a keen understanding of how to make the system work. She was a success.

Hillary, on the other hand, created a chaotic bureaucracy just to draft her health care bill, which ran to more than one thousand pages. She alienated members of Congress—even in her own party—as well as health professionals and the press. The collapse of her reform plan was a colossal personal and professional failure on her first national public stage. Her reputation was salvaged only by her grace during the Lewinsky scandal and her enthusiastic willingness to campaign and raise funds for Democratic candidates all over the country. And, once she had rehabilitated herself, it was still her alignment with Bill Clinton that led her to the next rung in her career: a Senate seat from the state of New York.

Condi's and Hillary's respective reputations in politics, too, were diametrically opposed. Condoleezza Rice has never been involved in personal or professional wrongdoing; Hillary has been embroiled in scandal after scandal, ever since she entered public life. She has always teetered on the ethical edge. Her unexplainable windfall in her commodities futures speculation; the circumstances of her Whitewater investment; the disappearance of her law firm's billing records; her role in the decapitation of the White House Travel Office employees; her solicitation and acceptance of personal gifts of expensive furniture, silver, and china during her last days in the White House while she was still first lady (but not yet a senator bound by rules about gifts); her acceptance of contributions and gifts from persons seeking presidential pardons; and the hiring of her brothers by drug dealers and others seeking pardons—all of these have led to the continuous cloud of doubt that has surrounded her personal and professional integrity.

Perhaps the most shocking example of her tin ear on ethical issues was her acceptance of furniture—and $70,000 in campaign contributions[7]—from Denise Rich, who was basically trying to buy a pardon for her fugitive ex-husband, Marc Rich. After a federal indictment charged Marc Rich with fifty-one counts of tax evasion and illegal trading with the enemy—Iran—during the hostage crisis of the late 1970s, Rich had fled to Switzerland and renounced his U.S. citizenship. In the wake of his ex-wife's gifts and campaign contributions to Mrs. Clinton and a $450,000 donation to the Clinton Library, Marc Rich was pardoned in the very last minutes of the Clinton presidency.

In stark contrast, Condi's past is without blemish. In her long journey among the elites of our power structure—first at home and then globally—she has never sought to profit personally from her position. It is difficult even to imagine her asking for expensive china, furniture, and silver from Hollywood stars, not to mention unsavory characters desperate for presidential pardons.

They are, indeed, a study in contrasts. If Hillary was tested in scandal, Condi won her spurs in war. As Hillary developed domestic policies, Condi mastered foreign affairs. While Hillary's candidacy is driven by enduring ambition, Condi's would be fueled by her own lasting achievements and experience in government service. Hillary has set her sights on a goal and pulled herself toward it. Condi has set her feet on her past and lifted herself above it.

Hillary's candidacy is obvious. Everything she does is calculated, carefully planned, and aimed at a White House run. With incredible cleverness and audacity, she successfully used her prestige as first lady to catapult herself into the Senate. Now she is using that Senate seat to jump back into 1600 Pennsylvania Avenue.

The prospect of Condi's candidacy is still obscure. As she rose to a pinnacle only one other woman before her had ever reached—the office of secretary of state—Americans have watched her style unfold on the international stage. They are getting to know her in real time, as she grows into her position and taps into her own possibilities. Her style has been described as "diplomatic activism." Every day she is seen on center stage all over the globe, promoting democracy by lecturing and cajoling our allies and standing tall against our adversaries. Her substance—but also her poise and elegance—are attracting attention and admiration. There is no sense that she is acting in a supporting role so that she can land the leading role later on.

No, while Hillary is always of tomorrow, Condi is uniquely of today. Echoing the Fleetwood Mac song that came to be an unofficial anthem for the 1992 Clinton campaign, Hillary never stops thinking about tomorrow. Each day is devoted to plotting, scheming, preparing, and positioning to advance further toward her goal. But Rice is fully the creature of today, fully involve in her current job, her current focus.

Of course, both women deny having any plans to run for president in 2008. In Hillary's case, the demur is traditional, usually couched in an often-repeated coy and calculated answer—"Right now I am focusing on being the best senator from New York that I can be"—rather than a flat-out rejection of the idea.

Condi's dismissals have been more emphatic. During an interview with the editors and reporters of the *Washington Times* in March 2005, she said she had no intention of running for president. A denial, but a soft one: "I have never wanted to run for anything,"[8] Rice said. "I don't think I even ran for class anything when I was in school." Her language, though at times it echoed Hillary's, was notably more modest: "I'm going to try to be a really good secretary of state," she told the *Washington Times*. "I'm going to work really hard at it. I have enormous respect for people who do run for office. It's really hard for me to imagine myself in that role."

But when the reporters pressed Rice for a "Sherman pledge," the time-honored definitive refusal first uttered by General William Tecumseh Sherman, the Civil War hero ("If nominated, I refuse to run. If elected, I refuse to serve"),[9] Rice backed off with a chuckle, saying that such a statement did not seem "fair."

She went further on the following Sunday during an appearance on *Meet the Press*, perhaps after conferring with the communications people at the White House. As the *New York Times* reported, "Secretary of State Condoleezza Rice, pressed by several television program questioners on the latest gossip about her in Washington, said repeatedly Sunday that she had no intention to run for president, no plans to run and no interest in running. Just to be sure, she finally said she would not run for president."[10] But, as the newspaper also noted, her statements "did not rule out" a future candidacy.

So what does Condi really think? Obviously she cannot jump into a presidential campaign within weeks or even months after assuming the mantle of secretary of state. There's no evidence that she has ever harbored a real ambition to run for president, and even now she seems to consider the possibility of a candidacy remote at best. But that does not mean that she would turn aside, as General Sherman did, if circumstances should create a genuine demand for her candidacy, with strong support in the Republican Party.

Hillary, on the other hand, approaches her candidacy in a much more traditional, political way. Through long and careful planning, she has nurtured her deeply ingrained ambition toward a run in 2008. Her step-by-step preparations to run have made her nomination almost inevitable. She has a master plan, and she follows it religiously.

And always, looming like a Sequoia in her background, is Bill, advising her, promoting her candidacy, raising money, winning support, running interference with the media, attracting a crowd, and, most important, holding out the promise of another Clinton presidency, another "two for one." Four more years of a President Clinton. He will be there throughout the race and beyond, a large and indefinable presence. What of her lack of administrative experience? Bill will help her out. No foreign policy background? Bill will be there. Her crash-and-burn record in her foray into health care reform? Don't worry. Bill will be around to take the blame and point out how much she has grown.

She'll do what presidential candidates must always do. She'll cross the country raising funds, winning primaries, cajoling delegates, massaging the media. Her ascent will be programmed, piloted by the smoothly running Clinton machine. All her life has been aimed at this moment, this contest. After eight years in the White House and five in the Senate, she knows how important each fund-raiser is, how much each handshake and letter means to her supporters, and she will not fail them. She'll memorize her lines and avoid mistakes assiduously. All of her public statements will be carefully scripted. There will be no spontaneous public outbursts, no chance for embarrassment.

And, whenever the going gets rough, she'll take refuge behind her spokespeople and refuse to comment. She'll remain above it all. While she may occasionally give vent to her innate sarcasm and loud partisanship at closed events, the public Hillary will be demure, on message, and aggressively and visibly moderate. Only on extremely rare occasions, where she is comfortably surrounded by her most liberal supporters, will the shrill attack-dog rhetoric of the real Hillary escape from behind her new, muzzled public face. But these windows into the true Hillary will be scarce; her road to the White House depends on hiding her true self, and she will carefully restrain herself.

Condi's path, if she chooses to follow it, should be very different: It must be the logical outcome of success at her day job as secretary of state. It is there that she must prove herself, winning the plaudits that can open the way to the White House.

Hillary's campaign will go through the usual pre-candidacy period. She will spend busy months raising money, recruiting support, and building her organization in the party centers of California, New York, Washington, and Florida. Condi's preparation will involve journeys of a different sort, as she builds up the diplomatic momentum to face the challenges of her mission: Paris, London, Moscow, and Beijing. Condi will be a statesman, a leader, a representative of the United States to the rest of the world.

And while Hillary's candidacy will take her to the early primary states—Iowa, New Hampshire, South Carolina, Delaware, and Arizona—Rice's will take her to trials of a different sort in Iran, Syria, Russia, China, Sudan, North Korea, and the West Bank. Her debut in the national spotlight will be anything but programmed. It will involve living by her wits, responding to changing circumstances, and showing herself to be a master craftsperson at this exalted level of global diplomacy.

But never has a secretary of state faced so many difficult tasks at once and yet been so well positioned to overcome them. The consistency of American might and determination in the caves of Afghanistan and the streets of Iraq has made the success of diplomacy—and democracy—much more feasible. It is the insight of the Bush-Rice foreign policy to reinterpret Carl von Clausewitz's dictum, "War is the continuation of policy by other means."[11] For this new secretary of state, diplomacy is war by other means. The American military mission in Iraq—and the sacrifice of so many good young men and women—has made diplomacy workable.

As Rice proceeds on her diplomatic odyssey, overcoming all the trials the gods of chaos can put in her path, she will look more and more like a possible president.

The process seems already to have started.[12] Robert D. Blackwill, U.S. ambassador to India from 2001 to 2003, recently observed in the *Wall Street Journal* that "diplomacy is flourishing once more at the State Department" under Rice. Citing the increasing alignment of the United States and India, the increase in international support for American efforts in Iraq, the revival of six-party talks in Korea, the unity of the West in demanding that Iran not to make its own nuclear fuel, and Israel's withdrawal from Gaza, he notes that Rice is already having an impact on American diplomatic fortunes after only a few months on the job.

The fact that Condi has not laid out a plan to run for president does not, by any means, signify that she won't run. It's not that simple. Compared with Hillary, she merely approaches her future in a very different way. She has never planned her own advancement with the same degree of precision that Hillary has. She hasn't had to. Her obvious talent has stood out among her peers, and her rapid promotions have always been the result. She needed no secret strategy or plot to ascend to the National Security Council under former President Bush, to the position of national security advisor to the current president, and now to the position of secretary of state. She was a woman on a mission, all right, but one with a substantive purpose, not a personal agenda.

Hillary is different. She is a plodder; she approaches the presidential race like a long to-do list. For her, the path to the West Wing in 2008 is already laid. The strategy is in place, the players on the team. For the past fifteen years, the Clintons have systematically built up a network of wealthy donors, influential supporters, and opinion leaders throughout the country, creating a Rolodex of millions. Like old-time ward heelers, they used the power of the presidency to reward these people by appointing them to jobs and commissions. They also understood the allure of invitations to the White House and used events like state dinners and Christmas parties to solidify the loyalty of their stalwarts. In their post–White House years, they've invited their A-list people to the Clinton Library and to Chappaqua and Georgetown for fund-raisers. They've stroked the backs of the key people in each state. Together, the Clintons still control almost all the levers of power in the Democratic Party. They know all the activists, and they work hard to keep them happy.

Recently, Hillary has been particularly focused on courting elected Democratic officials who will be automatic delegates to the 2008 convention. (They don't need to run in primaries to get seated.) Since her debut as a Senate candidate in 2000, Hillary has held more than three hundred fund-raisers for other Democratic senators, congressmen, and governors, collecting IOUs that she can redeem for votes on the convention floor in 2008.

And Hillary has the ground troops—the fund-raisers, the spin doctors, the speechwriters, the schedulers, the political handlers. Like the old Kennedy horses who answered the fire bell as it sounded, sequentially, for each new family member who entered the political fray, the Clintonistas are there waiting to help, anxious to ingratiate themselves with the Clintons and be part of their own version of Camelot. The games have already begun.

Under Bill's tutelage, but with the discipline he lacks, Hillary will scrupulously follow their jointly developed plan to recapture power. They may not spend much time together, but they are united on their journey back to Pennsylvania Avenue. Hillary will absorb all the lessons her husband's history has to teach and dramatically and obviously move to the center. The Clintons have always understood that they cannot attract swing voters with a leftist agenda. So, for the campaign, Hillary will become a moderate—at least in public.

For Bill has taught her the power of cutting against the image of your own political party. In 1992, Bill ran as a "new Democrat," advocating capital punishment, backing a work requirement for welfare, pledging to balance the budget and pass a middle-class tax cut. When Clinton criticized black rap star Sister Souljah for seeming to advocate black violence, he distinguished himself from the tapestry of liberalism that had been the backdrop of the failed presidential candidacies of McGovern, Mondale, and Dukakis. Now, against the history of another failed liberal candidacy, she is set to emerge as the new Clinton, the new moderate savior of a Left-addicted party. Playing off the extreme liberalism of the new

Democratic National Committee chairman Howard Dean, Hillary will position herself as the voice of reason and centrism.

Many have wondered why Hillary would allow a radical like Dean to win the DNC chairmanship. For one answer, one may look back at how her husband worked to narrow the 1992 field of candidates, until the final contest was between Clinton and California's Governor Moonbeam, Jerry Brown. Against Brown, Bill looked moderate. Against Dean, Hillary looks reasonable.

But Hillary's newfound centrism focuses only on issues at the margins of our politics. She may attack sex on television or call for more values in public life, but when the chips are down, she votes like a solid liberal, backing her party more than 90 percent of the time. When some Democrats crossed the aisle to work with moderate Republicans to avoid filibusters and speed judicial confirmations, Hillary was not among them. She hunkered down with the Left, resisting all compromise. When the time came to render judgment on the Bush tax cut or on his social security reforms—the key domestic issues of his presidency—Hillary led the attack on the president. Her centrism is manifest only when the cameras are rolling, and the issues aren't very important.

Hillary wants to be seen as defying stereotypes of party ideology—and of gender as well. In a time when women are still suspect as wartime leaders, she has actively positioned herself as a hawk in the War on Terror. When liberal female senators find their commitment to fiscal austerity in doubt, Hillary headlines her husband's success in balancing the budget and attacks Bush's war-driven deficit spending.

But Condi's way to 2008 is totally different.

She has none of the presumptive-nominee aura that Hillary has working for her. Her viability as a contender for the 2008 nomination will depend on whatever successes she has as secretary of state. She will first be seen as plausible, then as desirable, and, finally, as voters see Hillary move to the fore, irresistible. In the end, it is not Condoleezza Rice who will come to the voters asking for the nomination, but they who will come to her, imploring her to run.

Money will be raised for Condi, but she won't be the one raising it. Support will flow to her, but it will not be through her importuning. For once a presidential candidate will not be a senator courting publicity, but a secretary of state solving problems throughout the world.

Can Rice be nominated? The vacuum in the Republican 2008 field makes it quite possible that she can. There is no heir apparent. Cheney's health isn't strong enough, and nobody else from the Cabinet stands out. Rudy Giuliani and John McCain are the early front-runners, gathering together more than four out of every five decided votes in the polls. But Rudy is too liberal to win the nomination. And McCain showed his limited appeal to GOP primary voters in 2000, when he won the votes of Independents but lost the vast majority of registered Republicans to Bush. As meritorious as these two men are, they aren't going to win the Republican nomination. They are the early front-runners, but they'll fade in the opening rounds.

Their likely demise will leave an enormous vacuum. Candidates like Bill Frist, George Allen, George Pataki, Mitt Romney, and Chuck Hagel will get serious attention. The dominant reaction of Republican primary voters to this cast of unknowns will be "Who?" There will be a search for a real candidate, someone of stature, someone charismatic who can beat Hillary. And the party faithful will turn to Condi Rice. Well known, well liked, and well tested, she will rise to the top of the heap even if she isn't actively running.

America has not seen a real draft of a presidential candidate since Dwight D. Eisenhower in 1952. Yet popular acclamation can be one of the highest expressions of democracy: George Washington himself was essentially drafted into the role of chief executive. Republican voters will draft Condi because they need her; she will attract their support, without meaning to, as she shows the nation what an international force she can become.

A draft is especially possible at this time in our history, for—as the 2004 election results revealed—there has been a seminal change in U.S. politics. That was the year that the political ruling class was turned upside down: The opinion leaders and journalistic elite became the followers, while the mass populace—frothing with political interest and activity—emerged to take its place at the forefront of political change. Politics is no longer a spectator sport. Instead, it has become America's foremost hobby. An informed electorate is actively and regularly participating in the electoral process in innovative ways.

On the left and on the right, ordinary people found themselves in the vortex of the national campaign in 2004, each battling to be heard, outshouting the mainstream media and creating in the process a new, lower center of gravity for our politics. It's just the kind of environment in which the grassroots activists can decide who they want to be president—and go out and get her into the race.

This grassroots domination of politics in 2004 began when the Internet impelled Howard Dean upward so far and so fast that he almost beat John Kerry for the nomination. Then, when Kerry decided to build the edifice of his candidacy on the shaky foundation of his Vietnam record, the Swift Boat veterans, with very little money and no political experience, bested the Democratic publicity machine and brought the truth to the voters. And when CBS News and Dan Rather smeared the president's National Guard record, it was the bloggers who exposed the forgery on which their report was based. Finally, it was the 1.6 million Republican workers—and their Democratic counterparts—who brought out twelve million more votes for Bush on Election Day than he got in 2000, and nine million more for Kerry than Al Gore received four years earlier. America had never seen anything like it in the entire twentieth century. One had to go back to the hurrah politics of Andrew Jackson and "Tippecanoe and Tyler Too" to find an equal for the street-level politics of 2004.

If the street beat Kerry, it can also nominate Rice. The same avalanche of individual activists, each doing their own thing, can animate the draft-Condi movement. So widespread is the admiration for this self-made woman and so ubiquitous the fear of the Hillary juggernaut, that it may well be the spontaneous outpouring of hundreds of thousands of people that could propel a Rice candidacy.

As Condi distinguishes herself with her performance as secretary of state, her growing flock of supporters will come together online until they have reached critical mass, raising funds, generating volunteers, e-mailing friends, blogging, and reaching out to neighbors to fashion a real grassroots organization for her. As the political season approaches, volunteer groups will spring up in the early primary states, gathering signatures to put her name on the ballot. Condi herself need not endorse the movement. She simply needs to avoid giving a definitive no as these efforts gather force around her.

Each month, new national public opinion polls will show her to be gaining ground among Republican nominees. While the declared candidates battle it out, slinging negatives at one another, Condi can rise to the top precisely by abjuring the artificial fray of U.S. politics and earning her credits by mastering the real tests of international crisis and diplomacy.

It almost happened once before. In the autumn of 1995, General Colin Powell, newly resplendent in his post–Gulf War prestige, published his memoirs just as the pre-primary process for the 1996 Republican nomination to oppose Clinton was gathering steam. Inside the White House, Clinton was panicked. He ranted and railed apoplectically, to all within earshot, that Powell didn't deserve the "free ride" the media was giving him. "They won't ask him tough questions," Clinton would complain loudly. "They're all guilty white liberals and they want to use him to beat me," he shouted.

For a while, Powell seemed unstoppable. As he careened from one packed book signing to the next, his name soared to the top of all the presidential polls. Enigmatically, he refused to acknowledge the political firestorm around him and would not address the possibility that he might run in 1996. "It's the modern man-on-horseback," Clinton complained, drawing a comparison to the generals who would periodically stage coups in Latin America. He worried about how to run against a phantom, a creation of popularity, rather than the product of a conventional political surge.

Then came the bad news: Powell couldn't beat Dole in a Republican primary. His support for affirmative action, gun control, and an array of liberal positions undermined him and left him without a party. "Congratulations," I told Clinton after showing him the poll demonstrating that Powell wouldn't get the nomination—and therefore, I said, would not run. "You just won the election." Clinton stared blankly at me, doing the math in his head, then nodded.

But Condi is not Colin. And 2008 is not the same as 1996. Back then, Powell had to live off the residual legacy of his Gulf War achievements. But Condi will find her inadvertent candidacy fueled by her own real-time accomplishments on the world stage. She will demonstrate her ability to be president by acting out part of the job before a global audience. Her accomplishments will be current, vivid, and part of a growing legacy. Powell's blitz happened when his fame was on the downtrend from the Gulf War. Condi's will come on the upswing.

And wouldn't a Condoleezza Rice candidacy change America? The very fact that an African American woman could actually become president would send a powerful message to every minority child that there is no more ceiling, no more limit for black Americans in elective politics. The sky would now be the limit.

And the national stain that began to spread throughout our land when the first slaves landed at Jamestown, Virginia, would be erased. Condi's election would be the last battle of the Civil War, the last civil rights demonstration, the end of a saga that has haunted us since our nation was born. In a land where the signs once read "No Irish need apply," wasn't the election of John F. Kennedy the death knell of anti-Catholic bigotry? When he sat down after giving the most inspirational inaugural address since FDR's, you could feel the prejudice recede.

Racism remains one of the most fundamental problems of our nation. Its scars are so deep that they often have threatened to rip us apart. What greater social good could there be than its eradication—and what better way to do it than to elect a black woman as our next president?

2

President Hillary: How It Could Happen

Make no mistake about it. If the next presidential election were held today, Hillary Clinton would be in your face, exuberantly delivering her victory speech on every television network and beginning the redecoration of the White House, starting with the designation of the office for her chief adviser and the new first husband, Bill Clinton. (His would be the one right next to the president's dining room—the one with the small eye-level window in the door, so she can easily see what he's up to.)

Hillary is hot. She's popular. She's confident—and with good reason. She is, by far, the Democrats' top choice, and she has the support of women voters—the key swing group who make the difference in American elections. Money, for her, is no problem. Her donors love her and don't mind giving, over and over again. Her plan to win the nomination is viable, and she never wavers from it. She's built a loyal team, strategically placing her former staffers in positions at her various political committees as well as in the National Democratic Party. She looks great—the days of crazy hairdos and wacky clothes are long behind her. Everything is clicking just right; barring yet another Clinton scandal, she looks unbeatable against the regularly mentioned field of Republican candidates. She's a winner, and she knows it.

Hillary has found her groove. Her message is tight, clear, and extremely controlled. It reads: *Hillary Clinton is a hardworking, effective moderate who can collaborate with even the most conservative Republicans on joint, highly visible (and usually uncontroversial) projects. She's highly supportive of the military, capable in foreign affairs, and fighting to keep pornography and violence away from children. She's experienced: She spent eight years in the White House. She's independent of her husband, although very much married, and she's serious. She is NOT—repeat, NOT—A LIBERAL.*

Through her carefully scripted statements to the press and prominent photo opportunities, she underscores her new image as the smiling nonpartisan dealmaker. Here's Hillary with Rick Santorum, talking about children's television. There's Hillary with Newt Gingrich, discussing the privacy of health care records. Watch Hillary and Bill Frist as they agree on health care technology. Day by day, she scours the list of her right-wing colleagues to find areas of mutual interest and joint publicity. She's a whirlwind of legislative walk and talk. No matter that she rarely actually passes any bills herself. She is convinced that it's only the message that counts, only the message that is remembered. And she is broadcasting hers, loud and clear.

She uses the media to bolster her image as a player in foreign affairs and defense policy, and they never point out her lack of credentials. Hillary's recently acquired seat on the Senate Armed Services Committee has given her a platform, but, so far, she has not been influential in any matter of importance. When it comes to true accomplishment, Hillary is Foreign Affairs Lite. Her utter lack of grounding in foreign affairs has not stopped Hillary from pretending to be at the center of all things diplomatic. Her amateur musings on the events of the day—in the Middle East, in Iraq, in Africa—are translated into serious press releases. At best, she merely exploits the relationships her husband forged with foreign leaders while he was president. Like a shadow secretary of state, she welcomes friendly visiting dignitaries on their way to Washington and meets with representatives from countries like Canada, Israel, and Ireland on their way to Washington. She can do nothing for them, but as a courtesy—and because one day soon she may be president—she is courted by foreign leaders. She's the first Democrat they come to see. On the surface, her press statements may concern Syria, Iran, or Iraq. But they all share a common subtext: She's in the game. She's declared herself a principal in world affairs, and no one has challenged her. This is all despite the truth about her time at the White House—which is that, except for accompanying her husband on foreign trips, where she was given the separate first lady's tour of schools and hospitals, she did not participate in any matters of state during her husband's presidency and has no real experience or expertise in foreign affairs.

Hillary is incredibly disciplined. She moves carefully and never makes a spontaneous statement. And you'll never see photos of Hillary with Ted Kennedy or Howard Dean, talking about her support for the filibuster or her position on euthanasia for Terri Schiavo. Or with Barbara Boxer talking about late-term abortion. That would be way off-message. That's for the old Hillary. No, for the moment at least, she must be the new Hillary at all times. The old one will remain in mothballs until she recaptures the White House and goes back to being a liberal. Her old friends know this—and they can wait.

There are those who take Hillary Clinton lightly. This wishful thinking is tempting, but ultimately dangerous. Many people, especially conservatives, simply cannot bring themselves to believe that any sane person could support her for president. They remember vividly all of his scandals and hers; they recall her reinventions and her ideology, and they assume that other people feel the same way they do. Their personal distaste for her so overwhelms their perspective that they are blinded to reality, and their otherwise sound judgment becomes highly distorted. They truly don't believe that Hillary could ever pull together the national support she would need to be elected president. In due time, these conservatives believe, voters will easily see through her and reject her candidacy.

But that's just what a lot of people thought—and hoped—when Hillary ran for Senate in 2000. They did not believe—they could not even imagine—that savvy New Yorkers would fall for Hillary Clinton. A carpetbagger who had never held public office, who had never even lived in New York, who was suddenly claiming to have Jewish ancestors and a New York Yankees heart?

Those people, of course, were wrong. Hillary won the Senate seat by twelve points, winning upstate as well as downstate. She charmed the skeptics and hoodwinked the media, turning the usually skeptical fourth estate into supportive lapdogs.

Her tactics were audacious. Using the sympathetic bump in her popularity after the Monica Lewinsky scandal and her considerable leverage as first lady, she knocked all of her possible Democratic competitors out of the race—and then used the extensive perks of the White House and endless favors of the presidency to captivate party leaders and raise the millions she needed to win.

Those who are still skeptical about Hillary's chances of becoming president need to reflect on her wide-ranging appeal. She is obviously popular among liberals, but here is a far more amazing statistic: In a May 2005 *USA Today*/CNN/Gallup poll, 33 percent of conservatives—yes, *conservatives*—said they would be likely to vote for Hillary for president.[1]

Do not underestimate this woman!

WHY HILLARY WILL WIN THE 2008 DEMOCRATIC NOMINATION

Hillary Clinton has a lock on the nomination of the Democratic Party. Forget Kerry. Forget Edwards. Forget Evan Bayh. Forget Howard Dean. Forget Al Gore. Forget a host of wannabes. It's going to be Hillary.

As always in politics, all you have to do is follow the money. It leads directly to Mrs. Clinton. Ever since she embarked on her independent political career, she has been a nonstop fund-raising machine. In her Senate race in 2000, she raised and spent $30 million.[2] But that was really just the beginning. From the day she was elected, Senator Clinton has simply never stopped raising money. There is a dual purpose to this obsession: to help Democratic candidates and to expand her own base of supporters and donors.

The money she raises is not just for her own campaigns. She is very generous with her time and raises money tirelessly for other Democratic campaigns for Congress and governor. She raised money for the Kerry presidential campaign. She raises money for her own Political Action Committee, which she then donates to other candidates. She raises money for the DNC, state parties, and local candidates.

But this is no selfless task. Every dollar she raises for others only serves to increase her own political and financial capital. Hillary enjoys a symbiotic relationship with the all of the candidates and committees she supports. She brings a star quality to the events she hosts or attends and guarantees a large and generous crowd. That wins her the loyalty of the candidate, which can be useful down the line. And, at each stop, she adds names to her own list of backers and contributors.

In the election of 2002—the first after she became a senator—the *New York Times* described her as the "Democratic Party's single best fund-raising draw, the only one, other than her husband, who can pack a room . . . she has campaigned, or held fund-raisers, on behalf of more than 30 Democratic candidates for House and Senate who are in close races, as well as for three Washington-based Democratic campaign committees."[3] The newspaper noted that "her Washington home has become a conveyor belt of fund-raising dinners and receptions that Democratic candidates clamor to climb aboard."

After 2002, she actually accelerated her fund-raising—and not only for herself, but also for deserving Democrats from coast to coast. By now, according to the estimates of her own staff, she has raised at least $45 million for other Democrats since 2000.[4]

Now that Hillary is gearing up for her own reelection race in 2006 and for the presidential contest beyond, her fund-raising has revved up yet another notch.[5] She raised $4 million for her Senate campaign in the first quarter of 2005 and had $9 million on hand as of March 31, 2005.

But Hillary's fund-raising efforts are dwarfed by those of her operatives. The Clinton people, sooner than anyone else, mastered the art of raising money in the new environment created by the passage of the McCain-Feingold campaign finance legislation in 2003. Under these new rules—perversely called "reforms"—political parties and candidates, per se, are sharply limited in the size of donations that they may accept to fund campaigns. Now, no person can give more than two thousand dollars to any one candidate. The days when nominees and their parties could rake in huge "soft money" donations and spend them on their campaigns are over.

Even as McCain-Feingold was closing one loophole, however, the legislation was creating another—this one big enough to pass an ocean liner through. While neither candidates nor parties are permitted to take soft money, supposedly independent political committees, called 527s after the section of the law authorizing the loophole, now can accept it in any amount.

Harold Ickes, the Clintons' longtime fund-raising expert and thug-like political operative, quickly grasped that the McCain-Feingold campaign finance law had made the established political party committees secondary, and therefore obsolete, in the process of funding election campaigns. The Democratic and Republican national committee and the existing political action committees in both parties would no longer be the vehicle for the big money. From now on, the central role would go to soft-money political committees, newly exempt from the draconian limits imposed on the national party committees. While the restrictions made the party apparatus impotent, it left issue groups, labor unions, and informal ad hoc groups free to do as they wished.

In the years before McCain-Feingold, fund-raisers could count on the long-term institutional loyalty of the party faithful to the official fund-raising committees in Washington. When a fund-raising appeal came from the Democratic or Republican national committee, the party's supporters could open the envelope with confidence, knowing who was doing the asking and trusting them to spend the money wisely. But once the party was prohibited to ask for large checks, a new bond had to be forged between those raising the funds and the big donors who were willing and able to write the checks. The question was: Who would fill the void? Would labor unions step up their fund-raising efforts? Or would liberal special interest groups, like the environmentally focused Sierra Club or the National Abortion Rights Action League (now called NARAL Pro-Choice America), collect more money?

Hillary and Harold were determined to step into the breach. To let groups with an outside or substantive agenda replace the parties as the money-raising machines was anathema to the Clinton machine. The donors had to be kept in the family.

So, as soon as the 2003–2004 election cycle began, Harold Ickes—who has been at the fringes of many of the Clinton fund-raising scandals—established a new soft-money store, with a new shingle hanging outside. No longer did he ask donors to write $100,000 checks to the Democratic Party or to Hillary's campaign committee. Instead they set up a 527 political committee called Americans Coming Together (ACT) and deposited the same huge checks in this new bank account.

With Ickes at its helm, ACT became a vehicle that was loyal to Hillary and would set about winning the trust of the big donors vital to her future career. Ickes, who had served as Bill's 1992 de facto campaign manager and as deputy chief of staff in the Clinton White House, is Hillary's chief political operative, adviser, confidante, and aide. At ACT, Harold did what he does best—using loopholes in the campaign finance law to raise piles of money.

Realizing that in the new world of political campaign finance, the list was all-important, Harold Ickes set about fashioning a base of donors to ACT that would last long after the Kerry campaign became simply the answer to a Trivial Pursuit question.

Ickes treated the 2004 elections as spring training for the real job—putting Hillary in the White House again. While nominally raising funds for the lackluster campaign of Kerry and Edwards, he was, in fact, preparing for the role he will play in 2008 raking in money for Hillary for President.

Harold formed a marriage of convenience with George Soros and convinced him to donate at least $7.5 million to Americans Coming Together.[6] Soros, a Hungarian immigrant who fled Hitler's terror, is a dedicated partisan and backer of liberal causes. He abhors the GOP and has spent much of the past two years attacking President Bush. And, unlike many of the very, very rich, he is prepared to donate what it takes to win elections for his causes. He regularly intervenes in foreign election campaigns, using his fortune to undercut regimes not to his liking and is blamed for starting the entire East Asian currency crisis by allegedly staging a raid on the Malaysian ringgit. He has clout and he knows how to use it. Now, for Ickes—and therefore Hillary—he has become a not-so-secret weapon.

Ickes blazed a path through the thicket of the new campaign finance rules. ACT raised $137 million in the 2003–2004 campaign cycle. This is likely only a fraction of what it, or its replacement, will spend when Hillary—the real deal—runs for president. Some dry run!

ACT fell apart after the election but Ickes had established a fundraising mechanism for soft money that had won the trust and support of thousands of big donors. No longer would the Clintons have to rely on special interest groups or party committees to pay for their campaigns: Ickes had reinvented the wheel, but this time it was designed to carry the Clinton bandwagon to victory.

Congress, of course, needs to pass a real campaign finance reform bill that eliminates all soft money from anywhere in American political campaigns. But it won't happen—not in this lifetime. The most we can expect from the big money boys . . . and girls . . . on Capitol Hill is legislation banning 527s and opening up a new loophole to replace it.

Just add up the numbers: ACT raised $137 million in the 2004 election. Hillary's PACs and other groups raked in $45 million more that was given to other candidates, and her own campaign committee brought in another $9 million. Hillary and her operatives have raised almost $200 million since her election to the Senate, a breathtaking amount that makes the fund-raising of any other U.S. politician (apart from the president) seem minuscule by comparison. Welcome to the big leagues.

And Hillary knew how to use the money. She set about buying the support of convention delegates by raising money for them. Today, years before the 2008 nominating process swings into gear, one delegate in five to the Democratic National Convention has *already* been selected. Before the primaries. Before the caucuses. These are the super-delegates—Democratic congressmen, senators, and state party chairmen. These luminaries are entitled to the same one vote as any elected delegate who runs in the primaries. But, unlike the primary delegates, they are not instructed by primary voters to support a specific candidate. They can vote for whom they please. And Hillary showers them with campaign funds to assure their loyalty.

What are the super-delegates, and how did they come to be?

After the debacle caused by the domination of the 1968 Democratic National Convention in Chicago by the party bosses—which culminated in the spectacle of police battling dissident Democratic students in the streets—both parties reformed their nominating procedures to make them more democratic and responsive to the will of the voters. Henceforth, every candidate had to run in primaries or caucuses and would get only those delegates the voters actually awarded them.

But a by-product of this democratization was that the party bigwigs found themselves totally excluded from the nominating process of their own party. Unless they backed the right horse, they were out. Senators, governors, congressmen, even party chairmen found themselves locked out of the 1972 Democratic Convention—and they didn't take it very well.

When Chicago mayor Richard Daley, who had ordered the police mayhem in 1968, tried to get onto the floor of the 1972 party conclave, he was physically escorted out of the hall by the ushers. McGovern, the eventual nominee, knew this was no way to treat the man who was supposed to deliver Illinois to his ticket. But McGovern learned of the lockout too late to reverse it. He lost Illinois—and forty-eight other states.

After the 1972 disaster, the Democratic Party amended its rules once again. This time, all national elected Democratic officials and state party leaders became ex-officio delegates, with a full vote each. As a result, these super-delegates, who make up 20 percent of the total delegates, can propel a presidential candidate 40 percent of the way to a majority at the convention before the first primary is ever held.

And most of them will be for Hillary. Why? Because she has already bought and paid for them! All those fund-raisers at her Washington home for Democratic candidates throughout the country have a point: They are designed to win the votes of the super-delegates and provide an early advantage to Hillary. She knows that every vote counts, and she has already begun the tally.

But it is not just the party elite, with their super-delegate power, that love Hillary. The entire Democratic Party base loves her. Her memoir, *Living History,* was reported by BookScan, which keeps an unofficial tally on nationwide book sales, to have sold at least 1,324,727 copies.[7] Most publishers add another 30 percent above the BookScan figure to derive the definitive total sales of a hardcover book. Based on the numbers, almost two million people bought Hillary's book.

Her book tour resembled the tour of a chart-topping rock singer, hitting all the major U.S. cities. The lines of fans extended around the block at each bookstore where she appeared. For months, wherever Hillary went, mobs clamored to buy her book, shake her hand, and get her autograph. No other presidential candidate since Bobby Kennedy—not even her husband—has attracted such a following.

Hillary had three goals in publishing her memoirs. First and foremost, she wanted to make money. She was tired of living on a government salary—even the $200,000, plus housing and perks, that a sitting president collected.[8] She frequently pointed out that she and Bill didn't even have a home of their own, as if that were someone else's fault. Throughout Bill's terms as governor of Arkansas, she complained bitterly and loudly about the "sacrifices" that she and Bill had made in order for him to be president. Some would argue that living in the most luxurious home in the United States, with stunning antiques, priceless art, gorgeous flowers, world-class chefs, and a staff for the residence alone of more than five hundred people, was scarcely a hardship. Add to that private jets and helicopters; limousines; a weekend resort in the mountains; free health care insurance and round-the-clock medical care; unlimited entertainment of family, friends, and donors; free access to first-run movies; endless untaxed gifts of clothing, furniture, china, silver, vacations, and travel; and it sounds like a pretty good deal. But even this lavish lifestyle, to which Mrs. Clinton had grown quite accustomed, wasn't enough. And, besides, it was all coming to an end.

Writing her memoirs easily eliminated that problem. Simon and Schuster, her publisher, agreed to pay her $8 million advance between Election Day, November 7, 2000, and her swearing in on January 2, 2001—in that narrow, critical window of time after she won the Senate race and before her January swearing in, when ethical restrictions about accepting an advance would kick in. According to published reports, she was initially offered only $5 million on the first day of the auction she held during her final weeks in the president's house. Before the auction, publishers were invited to meet with Hillary in the White House, that storied mansion where Abraham Lincoln and Franklin Roosevelt tracked troops and battles. How history has changed! Hillary's second goal was to tell her story in a way that emphasized what she wanted and ignored what she chose to erase from her biography.[9] As we've noted elsewhere, deciding to call her book *Living History* took a certain amount of chutzpah: In fact, Hillary reinvented and rewrote history in the book to present herself in a favorable light, assail her critics, and portray herself as the new Hillary, one who would be attractive and sympathetic to the voters. The book is full of folksy stories about her happy family life and marriage, as well as exaggerations of her role in the Clinton administration. Her breathless narration of her utter shock at learning the truth about Monica Lewinsky—months after the country all understood it—is almost comical. But, here, too, she met her goal. Hillary laughed all the way to the bank and to a huge rise in public opinion polls. Millions of Americans, especially women, believed her stories and admired and respected her all the more.

Hillary's final goal was to use the book itself and the attendant publicity to rehabilitate herself, to change and improve her image—and, if that worked, to posture herself as a presidential candidate for 2008. And boy, did it work!

Before the publication of *Living History,* Hillary's popularity had fallen. Her poll numbers were significantly lower than they are now; some even had her favorability rating below 40 percent.[10]

Her popularity had been sagging under the weight of a growing Clinton fatigue. The gift and pardon scandals darkened the Clintons' departure from the White House. The voters were once again wary of her and her husband, believing both Clintons to be unethical and untrustworthy. Only days after she began her Senate career, Hillary was forced to hold a press conference and defend—unsuccessfully—payments to both of her brothers by people her husband pardoned. But all of that has faded away in the glow of success: Five years later and two years after her book was published, she's back on top with a 53 percent favorability rating, according to a June 2005 Fox News survey.[11]

Hillary now runs far ahead of the rest of the Democratic pack, topping the field of possible candidates with 44 percent of the Democratic primary vote.[12] John Kerry lags behind at 17 percent, John Edwards at a mere 13 percent. Since Election Day 2004, Hillary has gained four points, while Kerry has lost four and Edwards two. Once she becomes a declared candidate, one can expect her numbers to soar even higher.

And Hillary's move to the center is working. She is living proof that you can, indeed, fool some of the people all of the time. Are Americans blind? Don't they see the obvious calculation in these laughable alliances—with their almost daily photos and press releases—between Hillary and the right wing?

No less an observer of the Clintons' political career than former house speaker Newt Gingrich has noted the success of Hillary's move to the center. "Senator Clinton," he told the American Society of Newspaper Editors in a speech on April 13, 2005, "is very competent, very professional, very intelligently moving toward the center, very shrewdly and effectively serving on the Armed Services Committee—the first New Yorker to serve on the modern Armed Services Committee since it was created in 1948. And I think any Republican who thinks she's going to be easy to beat has a total amnesia about the history of the Clintons."[13]

Newt himself was recently the ultimate prop in one of Hillary's staged press events, designed to showcase how accommodating she is and how effectively she can work with politicians from all over the political spectrum—even the far Right. One wonders what impels these seemingly sane and focused partisan politicians to participate in Hillary's ridiculous dramas. Why would Newt Gingrich take the time to cooperate in Hillary's scheme to pretend to move to the center? Does he need to appear with Hillary so that he, too, can appear to move to the center? In much the same way, Hillary has hijacked Senate Majority Leader Bill Frist, South Carolina's Republican senator Lindsey Graham, and Pennsylvania senator Rick Santorum, a member of the Senate GOP leadership team, to appear with her in support of various pieces of relatively uncontroversial legislation. During the same time period, she also taped a gushing video message to be delivered at a roast of her former nemesis, New York's former senator Alfonse D'Amato, the chairman of the Senate Committee that investigated her Whitewater dealings. In the video, she said that she was proud to call him a friend.

Excuse me, am I hearing right? Does Hillary think she's running for the *Republican* nomination? At this rate, it won't be long before we can expect to see Donald Rumsfeld and Hillary in matching camouflage and helmets, happily climbing into a tank in Iraq together to encourage enlistees of both genders to sign up in the military. After that, she and Jerry Falwell might do a tour of high schools to promote abstinence.

But Hillary's strategy does not end with a move to the center and an embrace of the military. She has a multipurpose ace in the hole: her husband, the former president of the United States, Bill Clinton.

Many people still ask: Why does she stay with him? Obviously, there are many personal reasons. Beyond that, there are political reasons. First, together they have been a winning team for more than thirty years. It works for them. Even if they don't see each other very often, they still share two important common goals. The first is to elect Hillary as the first woman president. The second is to vindicate Bill's presidency.

Bill plays an enormous role in Hillary's quest for the Oval Office. Not only is he her major adviser, cheerleader, and fund-raiser, he is also a living reminder to the Democratic voters who adored him that he, too, would be back in the White House if she were elected. Without him, it would be very difficult for Hillary to be elected president. With him by her side, a third Clinton administration is within reach.

According to a June 2005 Fox News/Opinion Dynamics poll, 38 percent of voters would be "enthusiastic" about seeing Bill return to the White House as "first husband," while 33 percent would be scared. But most of that 38 percent vote in Democratic primaries![14]

In their curious relationship, the business partnership has always been an important component of their marriage. Because, for each of the Clintons, career comes first. The political content of their liaison is not the subtext of their marriage; it is the basis of their relationship.

After his withdrawal from the 1988 presidential race, when it seemed that Bill might never run for president, the Clintons began discussing divorce. And when their marriage was truly on the rocks, after the Lewinsky affair, it was Hillary's Senate ambitions that brought them back together. In a sense, it's politics that holds the Clintons together as they focus jointly on the project of their lives: nurturing their political ambitions to maturity.

Now that Hillary is running for president, it is Bill to whom she turns for advice and guidance. As Gingrich put it, Senator Clinton has "the smartest American politician as her adviser."[15] Already his fingerprints are apparent in the senator's transparent move to the center and her vigorous profile on foreign policy and defense issues. Having won the elections of 1992 and 1996 by tacking to the center, Bill is doubtless behind his doctrinaire liberal wife's move to the middle.

Bill also has Hillary talking about values. Again, this harkens back to 1996, when President Clinton responded to my polls showing that people suspected he didn't share their values by drastically changing his rhetoric. No longer were his public statements devoted to talk of programs, or budgets, or details. Instead he talked about values. Instead of speaking of the Head Start Program or the need to increase the number of children it reached, he talked about giving children more opportunity. We called this linguistic transition "teaching the president to speak Italian."

During the 2004 election, Clinton warned Kerry not to ignore the role of religion and values in the campaign.[16] According to *Newsweek*, even as the former president was about to undergo heart surgery, he urged Kerry to do more to appeal to mainstream American values in his campaign—advice the Democratic nominee didn't understand and didn't take.

Now, as Hillary speaks out about values issues like sex on television and parental notification of abortions for minors, she is doubtless hearing the same advice from her husband—counsel as sound in 2008 as it was in 2004.

But Bill Clinton's presence behind the scenes is not nearly as important as what he can do for Hillary in front of the TV cameras. Just as Hillary offset Bill's principal weakness by "standing by her man," so he can counter her chief problems by standing by his wife.

Any first-time candidate for president faces doubts about his administrative ability, foreign policy experience, and capacity to handle crisis. Particularly when the candidate is a woman who has held elective office for only a few years and who has no administrative or international experience, the doubts are likely to intensify.

Bill's presence assuages them; his experience reassures skeptical voters. The first husband would be live-in help. As Hillary navigates shoals unfamiliar to her, Bill will be at her elbow, helping her to avoid pitfalls. His record as president gives Hillary at least the reflected image of an incumbent's achievements to add luster to her campaign. His success at balancing the budget gives Hillary credentials to overcome the massive red ink into which the War on Terror, the Bush tax cuts, and the need to fund homeland security has plunged us. Bill's ability to avoid significant combat casualties while intervening in international crises in Haiti, Bosnia, and Kosovo lends credibility to Hillary's possible stewardship as commander-in-chief. Bill's moderation in signing the welfare reform act and strengthening the war on crime give those worried about Hillary's liberalism some ground for comfort—false though it may be.

Hillary has always been vague about how much or how little credit she is due for her husband's actions as president. While he was serving in office, she distanced herself as she sought to craft her independent image. But in running for the Senate, Hillary freely took credit for many of Bill's accomplishments. In *Living History* she is especially aggressive in overstating the role she played in his White House tenure, an account sharply at variance with the scant credit she receives for her inputs in Bill's version of the same events. (See our book *Because He Could* for an amusing comparison of what *she* says she did during his years as president compared with what *he* says she did.)

Now, however, the wraps are coming off. The former first lady constantly rattles off lists of accomplishments from the years of her husband's presidency, appropriating them as her own record, while contrasting them with what she considers the negatives of Bush's tenure. She contrasts Bill's robust record of job creation with Bush's anemic performance. She compares the years of balanced budgets in the Clinton White House with the swelling deficits of the Bush tenure.

Any aspiring presidential nominee from an opposition party would do the same thing. But Hillary's language always includes an implicit, unstated assumption: that they are, in part, her own. She deliberately fudges the boundaries between the Clintons to write up his achievements on her resume.

Finally, Bill Clinton is still Hillary's most useful surrogate in the political processes of fund-raising, parrying opposition attacks, and inveighing against her opponent. He is her heavy artillery, brought out when the going gets tough.

Who better to send to a fund-raiser to maximize contributions than Bill Clinton? Even if he no longer has the Lincoln bedroom or Camp David to auction off, he still packs plenty of star power. It will be Bill who will raise much of Hillary's money, just as he did when she ran for the Senate.

And who else could more effectively gut an adversary than Bill Clinton? When the former president of the United States attacks you, it is bound to hurt. And when he rebuts charges you have made against his wife, it packs a double punch: Not only was your adversary once the most powerful person in the land, he is also a husband standing by his embattled wife.

When it comes to winning the nomination, Bill Clinton will be Hillary's man, lining up delegates, schmoozing with party leaders, holding the super-delegates in line.

Bill Clinton clearly wants another term—or two—in the White House and will work hard to see that he gets them, through his wife. Recently, he has deliberately fueled speculation about her candidacy in numerous public comments about what a marvelous president she would be and how willing to help he would be.

And there is a secret urgency behind such statements: Beneath the Hillary brand lies a surprisingly shallow product. Even after serving for four years in the Senate, she has no notable legislative achievements to her name—nothing that would erase the disaster of her aborted attempt to reform America's health care system. She needs the reflected glory of the man who balanced the budget and reformed welfare to wipe out the unpleasant memory of what happened when she first arrived in the White House.

With that Clinton glow surrounding her, Hillary can do more than just win the Democratic Party nomination. She can be elected.

HOW HILLARY CAN WIN THE ELECTION

Look at the numbers.[17] George Bush got 62,040,606 votes on Election Day 2004. But his challenger John Kerry was close on his heels with 59,028,109. How many more votes could Hillary elicit from the Democratic base?

In evaluating such things, it's important to understand one thing: The Democratic Party is essentially a demographic grouping, while the GOP is an aggregation of like-minded people. Where Republicans rally around their standard-bearer because of what he and they believe, Democrats are inclined to belong to their party largely because of who they are.

In the 2004 race, for example, John Kerry drew heavily on the demographic base that undergirds his party.[18] He carried the African American, Latino, and single white women vote by large margins.

Together these three groups provided Kerry with more than half of his total vote, even though they accounted for only 38 percent of the total votes cast in the presidential election.

But Hillary will draw millions of new voters to her side that Kerry couldn't and didn't. Here's how.

In the presidential election of 2000, African Americans cast 10 percent of the vote.[19] In 2004, their share jumped to 12 percent.[20] When Hillary runs, their turnout is likely to swell even further.

While African Americans backed Kerry by a top-heavy margin in 2004, he was still essentially an acquired taste in the minority community. He had no great history with blacks; he was largely unknown among African Americans when he began his campaign for president and did little over the course of his campaign to change that. Blacks supported him for one reason: They didn't like Bush.

But African Americans know the Clintons well and like them a lot. Bill's genuine empathy for minority voters, his decision to locate his office in Harlem, and Hillary's focus on New York blacks during her Senate tenure will all encourage a much higher black turnout when a Clinton is on the ballot. How much higher? If black turnout should grow by a single percentage point in 2008, that would represent one million more votes for Hillary.

Hispanics voted for Gore by a margin of 62–35 in the 2000 election.[21] But Kerry carried Latinos by a bare nine points. Why the change? Because George W. Bush has worked overtime on appealing to the Hispanic vote. His eagerness to address them in Spanish, his sponsorship of a guest worker program for immigrants, and his Texas background all maximize his appeal to Hispanic voters. Bush also gained among the highly religious Catholic Hispanic vote by his opposition to gay marriage and his strong support for religious values. Remember also that Bush reversed the Republican Party's support for English-only public education and stopped cuts in school funding for the children of illegal immigrants.

But there is nobody with Bush's record on Hispanic concerns running for the Republican nomination in 2008. Neither Giuliani nor McCain nor Frist nor any of the other candidates would have the appeal that Bush has had for Hispanic voters (except, perhaps, Condi).

Hillary would likely bring back the Hispanic vote to the Democratic column in droves.[22] She runs very well among Latinos and carried New York's large Hispanic population by 76–24 in her 2000 Senate race. Hillary will likely carry Latinos by the same margin Al Gore did in 2000. But the turnout will be higher. In 2000, the rapidly growing ranks of American Hispanics cast only 6 percent of the vote for president.[23] But by 2004, their vote share had risen to 8 percent.[24] In 2008, we can probably expect another two-point rise in turnout, which would bring the Latino vote to a full 10 percent of the electorate. (Hispanics now account for 14 percent of the U.S. population.)[25]

If Hispanics account for 10 percent of the 2008 vote and go for Hillary by 65–35—as they did for Gore—she will get 2.3 million more votes from Latinos than Kerry got.

Now we come to Hillary's core strength—the female vote.

White women are the key to winning the presidency. They are the ultimate swing vote in American presidential politics.

White men (angry or not) are solidly Republican, and their votes are pretty much a given. In 2000, Bush beat Gore among white men by 60–36. The results among white men in 2004 were not much different: Bush beat Kerry by 62–37.

But it was Bush's gains among white women that reelected him. In 2000, Bush defeated Gore among white women by only one point—49–48.[26] But in 2004, he destroyed Kerry among white females by 55–44.[27] It was this ten-point swing in Bush's favor that gave him five million extra votes and cemented his victory.

The race for president in 2008 will come down to one simple question: Who can carry the votes of white American women?

Why are women so key? Because they vote in larger numbers than men do. In 2004, they cast 54 percent of the total vote. Last year, 8.4 million more women voted than men.[28]

Some basic facts:

- There are more women than men in the United States.
- Women are more likely to be registered to vote.[29] In the 2000 election, for example, there were 68.7 million women registered to vote compared with only 59.4 million men.
- Women who are registered are more likely to vote. In the 2000 election, for example, 56.2 percent of registered women voted, as opposed to 53.1 percent of registered men.

Until 1980, men and women voted in about the same proportions for each party. In 1968, for example, Republican Richard M. Nixon got 43 percent of the male and 43 percent of the female vote (although women were slightly more likely to vote for Hubert Humphrey, his Democratic challenger, while more men defected to racist George Wallace). In 1976, when Jimmy Carter squared off against President Gerald Ford, the Democrat got 51 percent of the men and 52 percent of the women.

But by 1980, when Ronald Reagan challenged President Carter, a gender gap had emerged. Carter actually carried women by 46–45, but lost because Reagan carried the male vote by 54–37 (the remainder voted for Independent candidate John Anderson). Since it appeared, the gender gap has never gone away. In every presidential contest since 1980, women have voted for the Democrat by between seven and twelve points more than men. Table 2.1 illustrates the sustaining impact of the gender gap.

The only time the gender gap shrank was in the 1992 election, when it dwindled to only four points. In that contest so many voters defected to Ross Perot's Independent candidacy that each party was reduced to its base voters, so the gender gap was less than usual.

The key point in these statistics is that the Democratic Party survives only because it draws disproportionately from women voters. Why?

Table 2.1
Gender Gap in Presidential Races

	Men	*Women*	*Gender Gap*
1976			
Carter (D)	51%	52%	1%
Ford (R)	48	48	0
1980			
Carter (D)	37	45	8
Reagan (R)	54	46	8
Anderson (I)	7	7	0
1984			
Mondale (D)	37	44	7
Reagan (R)	62	56	6
1988			
Dukakis (D)	41	49	8
Bush (R)	57	50	7
1992			
Clinton (D)	41	45	4
Bush (R)	38	37	1
Perot (I)	21	17	4
1996			
Clinton (D)	43	54	11
Dole (R)	44	38	6
Perot (I)	10	7	3
2000			
Gore (D)	42	54	12
Bush (R)	53	43	10
Nader (I)	3	2	1
2004			
Kerry (D)	41	51	10
Bush (R)	55	48	7

The timing of the emergence of the gender gap offers a good clue. Before the *Roe v. Wade* decision of the U.S. Supreme Court legalizing abortion in the first trimester of pregnancy, there was no real difference in how men and women voted. In 1976, the first election following *Roe,* the decision had little polarizing effect—because neither Carter nor Ford made an issue of abortion.

But in 1980, when Ronald Reagan consummated the marriage of the economic right and the religious right and spoke out frequently against abortion and for "life," a gap began to emerge between how men and women viewed the Republican Party. Other issues, including education, poverty, and peace, have also contributed to the difference between how men and women vote, but the abortion issue has led the pack.

American women vote dramatically differently depending on whether or not they are married. In 2004, married women backed Bush by eight points, while single women—whether widowed, divorced, or never married—voted overwhelmingly for Kerry by a 64–36 margin.[30] Interestingly, married men and single men tended to vote much more alike than their female counterparts did.

Since the gender gap is largely concentrated among single women, it seems likely that reproductive rights and issues related to the feminization of poverty have helped to account for its emergence. Single women have much lower incomes on average than their married sisters. So single women make up the vast majority of the people affected by the increases in the minimum wage—historically sponsored by the Democrats and opposed by the Republicans. The doctrinaire Republican opposition to abortion must, perforce, affect single women far more than it does those with husbands. The Democratic Party's historic focus on education and consumer issues probably also contributes to their historical advantage among women.

Since women who have never been married or are widowed or divorced account for 42 percent of all registered female voters, their preferences are quite important.[31] But single women are much less likely to vote than the general population. In the presidential race of 2004, only half of the unmarried women who were eligible voted—far lower than the 60 percent turnout overall. In 2004, twenty-one million unmarried women who were eligible to vote stayed home. If Hillary Clinton can attract even a small portion of them to the polls on Election Day, she can go a long way toward winning the election.

How will Hillary do among women voters? Obviously better than John Kerry. As might be expected, Hillary has a much more favorable image among women than among men. In New York State, where she has been most visible in recent years, she is rated favorably by women by a 67–30 margin, while her margin of positive ratings among men is only 53–41.[32] Undoubtedly, a feminist/female candidate would score much better than Kerry did among white women, particularly single white women.

Hillary's vote pickup among women is likely to be huge. If Hillary merely restores Gore's margin among white women (who cast 41 percent of all votes in 2004), she will pick up an extra 3.5 million votes. And, in all likelihood, she will gain even more.

So let's add it up: With one million more black votes, an increase of 2.7 million among Latinos, and 3.5 million more white women, Hillary stands to gain at least 7.2 million votes over John Kerry's take in 2004. And the numbers could prove far higher, particularly among female voters.

Will she lose any votes that Kerry got? Will white men vote against Hillary in larger numbers than they did against Kerry? It's not very likely. It is hard to imagine white men voting Republican by any more than the 62–37 margin by which they backed Bush in 2004. White men are already lopsided in their support for the GOP, and it's hard to conceive that they would reject Hillary Clinton by much more than they did John Kerry.

The Fox News/Opinion Dynamics polls confirm this supposition. When they matched Hillary against potential Republican rivals, the polls show her winning almost all of the Kerry vote.[33] John Kerry got the Democratic Party loyalists in toto to support his candidacy, but he didn't get too many other votes.

If Hillary betters Kerry's vote by between seven and eight million, she will easily surpass the margin by which Bush won the 2004 election. Bush beat Kerry by only 3,012,497 votes in his bid for a second term.[34] If Hiliary posts a gain of more than seven million votes, the Republican must come up with an extra four million or more just to keep up.

And in the most recent national polls, Hillary does very, very well. The May 2005 *USA Today*/CNN/Gallup Poll reported that 53 percent of poll respondents said "they'd probably support [Mrs.] Clinton in a run for president." Andrew Kohut, director of the nonpartisan Pew Research Center, said, "Over time, Clinton fatigue has dissipated . . . and people are looking back on the Clinton years more favorably."

Hillary remains polarizing, but her positives are growing, and her negatives are shrinking. The *USA Today*/CNN/Gallup Poll found that the percent of voters who were "very likely" to vote for Hillary for president rose over the past two years from 21 to 29 percent, while the number "not at all likely" to back her dropped from 44 percent in 2003 to 39 percent in 2005.[35]

The prospects of a President Hillary Clinton are growing stronger, month by month.

THE THREAT TO HILLARY: 2006

But before she reaches the 2008 presidential contest, Hillary Clinton must survive her 2006 race for reelection to the Senate from her adopted state of New York. Always a polarizing figure, she cannot take reelection for granted. Especially not if the right candidate emerges to challenge her.

Hillary enjoys very high ratings in New York, and easily defeats any likely challengers in most polls. Clearly she would fare very well against the typical conservative Republican. If the GOP wages its typical slash-and-burn negative campaign against her, Hillary will draw strength and energy from the backlash among New York voters.

Her 2000 GOP opponent, Rick Lazio, fell right into this trap. Instead of running a positive campaign stressing his own merits, this pro-environment, social moderate focused on telling people what they all already knew—that Hillary was a carpetbagger. Mrs. Clinton came right back and depicted her largely unknown opponent as a puppet of Newt Gingrich and the Republican Party. Since Lazio had not told voters about his real positions on key issues, Hillary was able to paint him as an ultra-right-wing nut whose election would imperil *Roe v. Wade* and turn the clock back on social progress.

But if Hillary's 2006 Senate opponent is pro-choice, pro-environment, and pro-affirmative action—and makes sure the electorate knows it—Hillary would be unable to hide behind negative campaigning and be forced to compare merits with her opponent. If Hillary's opponent were a woman with a moderate, pro-choice record, she could spell real trouble for the first-term senator.

And there is just such a person: Jeanine Pirro. The Republican district attorney of Westchester County, just north of New York City, has recently declared her candidacy for the race. The million residents of Westchester know Pirro as a tough prosecutor who has been especially innovative in formulating ways to prosecute pedophiles, Internet porn kings, and child abusers. Her proposal to strip teen drunk drivers of their licenses until they turn twenty-one will win her wide support. Pirro is just what Hillary fears most—a woman who stands for tough measures to protect children and who cannot be painted as a right-wing extremist. Since Jeanine agrees with Hillary on many of the litmus test issues—including abortion, affirmative action, gay civil unions, and more—she could be a real threat as an opponent to Hillary in 2006.

But Republican Senate candidates in New York State are often underfunded. In this Democratic state, most of the big money will line up behind Hillary. But Pirro may be able to count on a massive influx of funds from all around the country, as people who want to stop Senator Clinton in her tracks send in small—and not so small—donations.

And Hillary has a big, big problem in running for reelection to the Senate: She wants to be president.

Her ambition will prevent her from pledging to serve out her term as senator—six years—if she is reelected. As governor, Bill Clinton was able to get away with pledging to serve out the four-year term he won in 1990 as Arkansas governor and then breaking his word by running for president, but Arkansas is not New York. On this larger stage, a nationally prominent senator cannot get away with a broken promise as easily as the unknown governor of a small state.

This is not a minor issue for Hillary. Sixty percent of New Yorkers want Hillary to pledge "to serve the full six-year term" before she runs for reelection to the Senate in 2006, according to a recent Quinnipiac University poll.[36] In May 2005, on CNN's *Inside Politics,* Hillary was challenged directly on the issue: If "you were asked to pledge, at some point between now and next year, whether you will definitely fill out a six-year term in the Senate, what would you say?" Hillary refused to take the pledge. Instead, she said, "I am focused on winning reelection. That is what I work on every single day, just as I have worked my heart out for the last four years. And I'm going to continue doing that every day, and I'm not going to get diverted."[37]

Another recent poll, this time by Marist College, found that New Yorkers did not want Hillary to run for president in any event by a 51–44 margin.[38] Even 35 percent of the Democrats did not want her to run for the White House. Those who want Hillary to stay home are not necessarily Hillary haters. Many are simply New Yorkers who take her at her word—that she wants to be their senator!

Why this sensitivity to making sure Hillary serves out her term? Most voters in most states don't really care. Let the candidate do what he wants, they typically tell pollsters. But the special circumstances surrounding Hillary's adoption of New York as her home state and the way she first came there to run for the Senate may have made New Yorkers more sensitive to the issue.

Every New Yorker knew that Hillary was a carpetbagger. The closest she could came to claiming hometown roots was to say that she was part Jewish (her grandmother's second husband was Jewish—no genetic relation) and that she'd always been a New York Yankees fan! Most New Yorkers snickered at her clumsy attempts to identify with them, but they took her word that she, like so many others throughout the nation, liked New York and wanted to move there. The Empire State, after all, is filled with people from someplace else; it was hardly surprising to believe that Hillary might want to become part of the Big Apple. Many of the more gullible were thrilled that she wanted to represent New York and bonded with her in a personal way.

But if she jilts them and leaves New York behind like some sort of doormat or stepping stone, polls show that she may kindle popular resentment (much as the people of Arkansas must have felt). And if she should make it clear to New Yorkers that she is probably leaving, but that she wants them to reelect her to the Senate so she can continue to use them to advance her own ambitions, she may be surprised at the lukewarm reception she gets.

It is clear that the Clintons are aware of this threat to her reelection and are trying to spin it to avoid any erosion in her vote share. The tip-off? Bill Clinton has begun to muse publicly about the issue. In an early June interview with Larry King, he insisted that Hillary need not make any commitment to complete a second term. Why? Because other candidates in other states have refused to do so. His model? President George Bush, who refused to make a similar pledge when he ran for reelection as governor of Texas.[39] It's hard to imagine Hillary claiming George Bush as a role model, but when the shoe fits, she'll wear anything. In this case, though, the fit may not be as perfect as she hopes: Bush's refusal to promise a full term as governor obviously didn't bother people in Texas. The people of New York, on the other hand, have already made it clear that they want such a commitment from her. And if she refuses, a backlash is likely to follow.

This backlash would not be enough to get New York to go Republican in a presidential race, of course. But in the 2006 Senate race, it could be enough to persuade New Yorkers to consider a socially moderate woman Republican as a serious alternative.

After all, her opponent will be able to say: "I am the only candidate for senator from New York State who wants to *be* senator from New York State for the next six years."

All this raises a surprising question: Will Hillary actually run for reelection to the Senate? She says she will, of course. But does she really mean it? As long as she has no serious opposition, she will probably take the free ride and pocket an overwhelming reelection victory. But if she sees a moderate Republican woman coming at her, she may well want to get out of the way and not seek a second term. What has she to gain? Nothing. If she wins reelection, she merely keeps the seat she has. What has she to lose? Everything. A defeat would end her career. If Hillary starts to drop in the polls in New York State, she may get out of the Senate before she is embarrassed by the voters of New York. (Even a narrow victory would hurt her presidential chances.)

Terry McAuliffe, the former chairman of the DNC and the finance chairman for Clinton's 1996 reelection campaign, put it this way: "Why would Hillary take $10 million she could spend to run for president and waste it getting reelected to the Senate?"[40] But, $10 million is a very low figure for her race. She'll need about $40 million.

So why indeed?

3

How Condi Can Beat Hillary

Condoleezza Rice can defeat Hillary Rodham Clinton—and it may be that no other Republican can. Were she to run, her candidacy would strike directly at the three pillars of the Democratic Party's political base: African Americans, Hispanics, and white women.

The Democratic Party cannot win without fully tapping all three sources of votes. If African Americans had not voted solidly for Kerry in 2004, the Democrat would have been beaten in a landslide. The only reason the Massachusetts senator carried New York, California, Illinois, New Jersey, Michigan, and Pennsylvania was the solid support he got from African Americans.

The defection of Hispanics from the Democratic Party in 2004 had disastrous consequences for John Kerry, giving Bush most of the votes he needed to win.

And, in addition to African Americans and Hispanics, the Democratic Party is still dependent on white women for its political survival. It was John Kerry's failure to replicate Gore's performance among white women, more than any other factor, that caused his defeat.

A Hillary Clinton candidacy is particularly strong because of her appeal to all three bastions of Democratic power. Because of her husband's long identification with minority voters, her efforts to court Hispanic voters, and her own gender and record of feminism, she stands to cash in on the support of all three groups in a huge way.

But Condoleezza Rice, also a woman and an African American, blocks Hillary's built-in advantages.

How would Condi fare among blacks? Would she crack the solid phalanx of African American support for the Democratic Party, something no Republican has done in fifty years? A number of prominent black Democratic politicians think she could.

Bill Clinton's former secretary of agriculture, Mike Espy, the first black congressman from Mississippi—and a lifelong Democrat—thinks Condi would run well among America's blacks. Espy, who was one of two African Americans in Clinton's first cabinet, is still licking his wounds from the special prosecutor sicked on him by the Republicans. (Espy was found not guilty by a jury.)

"They are two brilliant women," Espy says, "evenly matched, both well rounded, both with interests outside of politics."[1] Contemplating the contest, he quips: "I'd love to buy a big bag of popcorn and watch them go at it."

How would the black community vote? "Their heads would be for Hillary," Espy predicts, "but their hearts would be with Condi." And which would they follow? "We often are emotional and follow our hearts. We would all like to have parents like Condi's—focused, encouraging, nurturing—and we'd all like to have a daughter like Condi," Espy says.

When I pressed him for a numerical prediction, the former congressman thought for a while and then said: "My guess is that the race [among African American voters] would be pretty much even. Hillary may have a bit of an edge because of the hegemony of the Democratic Party base, but Condi would run much, much better than any other Republican. My guess would be a 60–40 Hillary margin."

Sixty-forty! For a Republican to win four out of ten black votes would mean a major realignment in American politics. If Rice should realize anything close to such a gain in the African American vote—and do as well as Bush among the rest of the electorate—she would sweep to an overwhelming victory, a true landslide.

Espy is out of office, so he was willing to let us use his name alongside his predictions. When we spoke with another black Democratic congressman from a swing state who is still in office, he was more reluctant. Off the record, though, he went even farther than Espy, saying that he thought Condi would carry the black vote. "I think black America would be torn," he predicts, "but I think they would ultimately come down on the side of Condi Rice. Even with President Clinton campaigning for Hillary, I think Condi would represent a serious threat to Hillary. I'm for Mrs. Clinton. But I think Rice would be a very strong candidate."

And Jamal E. Watson, the executive editor of the largest African American newspaper in the nation, the *Amsterdam News,* believes that Rice would actually *defeat* Hillary among blacks by 60–40! Even though Watson's paper is located in the middle of Harlem, a few blocks from Bill Clinton's office, in the epicenter of Hillary's political base, he sees Rice as a "very strong candidate" among African Americans.[2]

If Rice were able to break even among black voters, she would switch 4.5 million votes from the Democratic to the Republican Party—a net gain of nine million in the margin between the two parties. Since Bush was elected by a 3.1 million-vote edge, the impact of so massive a switch is obvious. As Sean Hannity of Fox News puts it: "A ten-point swing [among blacks to the Republican Party] would be enough to win an election. A fifty-point swing would be incredible."[3]

But would blacks really desert their time-honored commitment to the Democratic Party? Massive African American migration from one party to the other has happened before. Prior to FDR, blacks voted overwhelmingly Republican. It was only in the 1930s and 1940s that they began supporting Democrats. Only when Johnson defeated Goldwater in 1964 did the African American vote become virtually unanimous in favor of the Democrats.

The history behind this is a classic lesson in political identity and realignment. After the Civil War, newly enfranchised Southern blacks (most African Americans lived in the South) voted Republican. The Democrats were the slaveholder party; the destroyer of American slavery—Abraham Lincoln—was a founder of the Republican Party.

In the decade after the end of the Civil War, it was the Republican Party that led the way in giving the newly liberated minority population equal rights. Overwhelming Republican majorities in Congress passed the Thirteenth Amendment outlawing slavery, the Fourteenth prohibiting discrimination, and the Fifteenth guaranteeing blacks the right to vote. The Democratic Party was strongly opposed to these measures and continued to advocate the cause of their Southern white constituents.

In the South, Republican committees in Congress combined with the Republican administration of former general Ulysses S. Grant to try to reconstruct the South by guaranteeing black equality and freedom. Backed by federal troops, northern Republicans flocked to the former Confederate states, imposing what whites there felt was a colonial regime. The largely white and implicitly racist history books I read in the New York City public schools of the 1950s and 1960s painted Reconstruction as an era of unbridled corruption and exploitation. Movies like *Gone with the Wind* popularized the cause of Southern whites disenfranchised by northern manipulators of their ignorant former slaves.

But the truth is that in the immediate aftermath of the Civil War, blacks had civil liberties, could own property, and had the right to vote—but only because the Republican Party guaranteed it with federal troops. The Democrats organized white resistance to civil rights through the Ku Klux Klan, which launched a reign of terror forcing blacks back into the margins of society by systematic lynchings and shootings.

As northern support for civil rights for blacks—never strong to begin with—began to wear thin, the Democratic Party began to come back in both the North and the South. Tired of footing the bill for a massive federal occupation of the South that seemed to go on endlessly, northern whites turned to the Democratic Party, which lured them in by saying it wanted to heal the wounds of the Civil War. But what the Democrats really meant was that it was time to abandon the Southern blacks to the ruthless terror of their former white masters.

When the Southern states were readmitted to the Union, their white majorities soon reasserted their power and propelled racist Democrats into office—often former Confederate officers and officials. When Grant left the presidency, all hope of securing rights for minorities in the South vanished; his Republican successor, Rutherford B. Hayes, was complicit in a deal to pull troops out of the South in return for the presidency.

Once the blue-uniformed federal soldiers withdrew, blacks quickly lost whatever rights they had, and an era of Jim Crow discrimination descended on the region. Those blacks who were able to vote, whether in the North or the South, backed the Republican Party while the Southern whites bloc-voted for the Democrats. This gave rise to the term "solid South."

Blacks remained loyal to the Republican Party until President Franklin Delano Roosevelt began to reach out to black voters. FDR had been elected without major black support; most African Americans voted for President Herbert Hoover—the last time a Republican would carry the black vote until Eisenhower in 1952.[4]

But Roosevelt's vigorous battle for the poor and against the Depression kindled enthusiasm among African American voters. More important was Eleanor Roosevelt's determined resistance to racial discrimination. When the Daughters of the American Revolution (DAR), then a reactionary, white-dominated group, refused to let black singer Marian Anderson perform at Constitution Hall, Eleanor organized a rally at the Lincoln Memorial, attended by

75,000 people, at which Ms. Anderson sang. As Michael Zak notes in his very important work *Back to Basics for the Republican Party,* Anderson began her appearance by singing: "My country 'tis of thee, sweet land of liberty, of thee I sing. . . ."[5]

Thus African Americans began their forty-year migration away from the Republican Party, giving a majority of their votes to Roosevelt in each of his three subsequent election victories and rallying to help reelect Democrat Harry Truman in 1948.

But the Republican Party remained very attractive to black voters throughout this period. When President Harry Truman proposed civil rights legislation, which triggered a walkout from his convention by the Southern wing of his party, it was because he feared that Republican presidential nominee Thomas E. Dewey would walk away with the African American vote.

In 1952 and 1956, the Republicans controlled the White House and regained their former ascendancy among black voters when General Dwight D. Eisenhower twice defeated Democrat Adlai E. Stevenson for the presidency. For the first time since 1928, a majority of blacks backed the Republican Party.

And Eisenhower gave them good reason to stay Republican. One of his first acts as president was to appoint California governor Earl Warren as chief justice of the U.S. Supreme Court. It was Warren who led the court to overthrow school segregation in its landmark 1954 decision in *Brown v. Board of Education of Topeka, Kansas.* As soon as the decision was handed down, Eisenhower ordered desegregation of the Washington, D.C., public schools; then, when Arkansas' Democratic governor, Orval Faubus, sought to block the court-ordered integration of Little Rock schools, the president sent in federal troops to enforce the judicial decision—while an eleven-year-old Bill Clinton watched in awe from Hot Springs, Arkansas.

In 1957 and 1959, Eisenhower and his progressive attorney general, Herbert Brownell, proposed strong civil rights bills to enforce the long-neglected Fifteenth Amendment and give Southern blacks the right to vote. Senate Southern Democrats filibustered the bills and succeeded in watering down their strongest provisions. And when the Southerners demanded that violators of the new civil rights bill have the right to jury trials (before all-white Southern juries), Democratic senator John F. Kennedy voted with the South, while Republican vice president Richard M. Nixon broke a tie in the Senate to kill the Southern amendment.

Back then, the most famous African American in the nation, Jackie Robinson, was a Republican.[6] And so was Condoleezza Rice's father. In her speech to the Republican National Convention in 2000, she remembered that "My father joined our party because the Democrats in Jim Crow Alabama of 1952 would not register him to vote. The Republicans did. I want you to know that my father has never forgotten that day, and neither have I."[7]

The Democratic domination of the African American vote really did not begin until 1960, when the Democratic presidential candidate, Senator John F. Kennedy, dramatically called Coretta Scott King, the wife of Dr. Martin Luther King Jr., after her husband was sent to prison in Georgia. While JFK expressed his sympathy to Mrs. King, his brother Bobby Kennedy worked the phones, and King was soon freed. It was the phone call heard round the nation, as Democratic campaign workers flooded black churches with word of the senator's intervention. On Election Day, blacks showed their appreciation by voting for Kennedy by a margin of 70–30, more than enough to give the Democrat the victory by a national margin of 49.7 to 49.5 percent over Republican Richard Nixon.[8]

In 1964, the black preference for the Democrats became a landslide, as President Lyndon B. Johnson rallied a grieving nation after Kennedy's assassination to demand passage of the strong civil rights bill JFK had proposed during his last year in office. Backed by a national outcry, Johnson jammed through the far-reaching legislation, which ended discrimination against blacks in virtually every area of national life. Ironically, it was only with strong Republican support that the bill was able to pass. Competing for support from black voters, Republicans, led by floor leader senator

Everett Dirksen of Illinois, pushed hard to terminate the Southern filibuster that threatened to kill the bill. In the House, 80 percent of the Republicans—but only 63 percent of the Democrats—voted yes, while in the Senate twenty-one Democrats (including Al Gore's father and current West Virginia senator Robert Byrd) voted no.[9] Only six Republicans joined them.

But one of the six naysayers was the Republican presidential nominee, Senator Barry Goldwater. In opposing this landmark legislation, Goldwater delivered a massive 94 percent of African American votes to the Democrats, and blacks have voted for Democrats ever since.[10]

Richard Nixon, having lost the black vote in 1960, made matters worse with his "Southern Strategy," which emphasized converting white Democrats from the South to the Republican Party by opposing school busing to achieve integration and inveighing against lawlessness—a thinly disguised code word for black urban ghetto rioters and criminals. Nixon's strategy succeeded brilliantly; by the 1970s the South was becoming a Republican stronghold, even as blacks were becoming solid Democratic supporters. The races had switched sides of the political divide.

But Bill Clinton changed the underlying political equation with his vigorous and successful efforts to cut crime and slash welfare rolls. By reducing these two causes of white angst, he attracted blue-collar voters—the so-called Reagan Democrats—back to the Democratic Party, without losing African American support.

The by-product of this easing of national racial tension has been that white racism has rapidly faded its one-time position at the center of the American political landscape. With crime, welfare, and unemployment down, the resentments that have historically undergirded bigotry and prejudice are no longer as virulent.

The strong interest in a Colin Powell candidacy in 1996 among the Republican grassroots—including much of the South—shows how much race relations have improved. Where once the courts had to gerrymander legislative districts to ensure the election of black congressmen by giving them "majority minority districts," recent court decisions overturning these artificial district lines have left most African American congressmen in office even as the minority populations in their districts have dropped. Whites are voting for black candidates in increasing and impressive numbers.

With racism on the wane, the chance for a rapprochement between the GOP and black voters is getting better and better. If the Republicans were to nominate Rice for president, the predictions of such esteemed Democrats as former Clinton agriculture secretary Mike Espy and his unnamed Democratic congressional colleague from a swing state might be very accurate.

If Condoleezza Rice can flip a significant portion of the African American vote to the Republican Party, she will truly perform a long-term service for the American people and politics. Why? Because—in a bitter irony—the truth is that, as soon as blacks got the vote in the wake of Johnson's 1965 voting rights act, their permanent alliance with the Democratics stripped them of any true political power.

Democrats treat the black vote as they would a golf handicap. As soon as the election campaign starts, party strategists automatically assume that they will win virtually the entire African American vote. In national campaigns, for example, they start with at least 10 of the 12 percent of the vote cast by blacks. It then becomes the task of the Democrats to win forty additional points from whites and Hispanics.

Republicans have the opposite perspective. No matter what the record of their candidate—or of the party—blacks never vote Republican in any significant numbers. The challenge for Republicans is to win enough votes among whites to offset the black vote for the Democrats. They never campaign among African Americans, except to convince white voters that they are not racist.

This bloc voting by African Americans is the last real vestige of a racial divide in American politics. Hispanics, who previously voted preponderantly for Democrats—usually by a 2–1 margin—have declared their independence from Democratic domination by voting for Kerry by only a nine-point margin in 2004.

Ever since their vote became competitive, Hispanic voters have been able to win important concessions from the Anglo-political establishment. Once the GOP favored English-only legislation; now the bulk of the party accepts bilingual education instead. Likewise, the Republican Party's historically hawkish policies on illegal immigration have ebbed as President Bush proposes legalizing Mexicans working in the United States as guest workers and allowing them to follow a path to citizenship. Under the Clinton administration, Republicans in California tried to pass a ballot proposition banning the children of illegal immigrants from free public schools (unless they were American citizens by virtue of being born here). But Bush has forced them to abandon their position and withdraw the legislation.

Even as both parties court Latinos, however, neither Democratic nor Republican politicians have paid much attention to African Americans, because they have traditionally been inert as a political base. In thirty years of political consulting, I have never been asked the following question: "How can I win the black vote?" Republicans usually asked me to poll only the white and Hispanic vote—not because they didn't like blacks, but because they knew that, no matter what their opinion was, they were going to vote Democratic in the last analysis anyway, and they didn't want their opinions to cloud the data.

This segregation of the black vote into the Democratic corral is deeply pernicious to our democracy. It eats away at the fiber of our freedom and creates an unchanging mass dedicated to one party regardless of policy, personality, or priorities. It has created a political atmosphere in which neither party cares much about the needs of the black community. The days when Democrats like Kennedy and Republicans like Eisenhower would vie with each other to address black demands—and thus win their support—are long gone. The black vote gets no attention from the political establishment precisely because it is not a moving piece of our politics.

But if Rice succeeds in putting the black vote up for grabs, it will end her race's isolation from the political process. She will reinvigorate the politics of black America permanently. Democrats and Republicans alike will compete for African American voters as they did in the 1940s, 1950s, and 1960s. She will make the entire American population—not just the white and Hispanic sectors—politically active again.

There is no greater contribution she can make to the future of American democracy.

Hillary Clinton is undoubtedly hoping for a large share of the African American vote in 2008, along with an increased turnout. But the key to her success has to be winning the women's vote by an overwhelming margin.

It was the switch of white women from Gore in 2000 to Bush in 2004 that made the president's reelection victory possible. White women will decide the election of 2008, just as they determined the outcome in 2004.

Obviously, Republicans would have a better chance to get the female vote if they run a woman for president, just as they would have a better shot at blacks if they ran an African American. But there is much about Condoleezza Rice, apart from her gender, which may endear her to women voters.

For one thing, she's single. It is unmarried women, more than their married sisters, who account for the bulk of the gender gap that favors the Democrats. While Rice is hardly the typical single mom, juggling kids and a job, she is an unmarried woman who made it in the world on her own. Single women are likely to relate to her biography more than they would to a woman whose success is intimately tied to her marriage and her husband's accomplishments. In fact, some single women may actually be jealous of someone who succeeds because a man powers them ahead; consequently, many unmarried women may find they have more in common with Condi than with Hillary.

Most studies show that two issues dominate the voting preferences of women voters: abortion and education.

On abortion, Rice is "mildly pro-choice."[11] While she opposes late-term abortions and Medicaid payments for abortion and wants parental notification and consent where a minor is involved, Rice supports the basic libertarian idea that it is up to a woman to decide whether to have an abortion, and she is against getting the government involved. So she is much more attractive to women than any of the other likely Republican candidates, except perhaps for former New York mayor Rudy Giuliani. None of the others who have been mentioned for the GOP presidential nomination is anything other than a doctrinaire supporter of the pro-life movement, up and down the line. Rice's position on this key issue is likely to be a big factor in making her more attractive to single women.

On education, Rice is an experienced educator, where Hillary merely writes about the subject. Condi's extensive experience at Stanford, as a teacher, a mentor, and an administrator, qualifies her to speak on the education issue in a way no other candidate in either party can. If Laura Bush's pedagogic background made her husband more attractive to women voters, imagine their reaction to a former full-time educator running for president.

Hillary, who has made education one of her central issues, has nothing to match Condi's credentials on the subject. Mrs. Clinton has never taught a class or run a school.

Obviously, only Condoleezza Rice offers the Republican Party a chance to close the gender gap—or, at the very least, to prevent it from widening further. If Condi could do nothing more than blunt Hillary's potential gains among female voters, she would be accomplishing a miracle. Hillary's attractiveness to women voters will be key to her chances of succeeding where John Kerry failed. If Condi neutralizes it, the senator's chances of victory drop precipitously.

Hillary might be able to survive a drop in her margin among blacks or women.

But how could she possibly survive both?

Condoleezza Rice is far more likely to win the support of both groups—and, in doing so, to reverse the identity politics that has hobbled the American electoral process for generations.

4

Being Condoleezza Rice

Shakespeare said it best in *Twelfth Night:* "Be not afraid of greatness. Some are born great. Some achieve greatness. And some have greatness thrust upon them."[1]

Throughout American history, we have had important leaders in each category. FDR, John F. Kennedy, Nelson Rockefeller, George W. Bush were all born great—fated by their families to play a prominent role. Lincoln and Jefferson, among others, achieved greatness. Only one national leader, George Washington, could be said to have had greatness thrust upon him.

Like most of us, Condoleezza Rice was certainly not born to greatness. Her great-great-grandparents were slaves. She grew up in segregated Birmingham, Alabama—one of the major flash points in the civil rights struggle that raged around her throughout her childhood years.

But neither has Rice striven single-mindedly, as Hillary has, to achieve greatness. Her style has not been one of ambition and grasping. Rather, as Nicholas Lemann observed in his excellent *New Yorker* profile of Rice, "several of the crucial turns in Rice's career have had the quality of an audition—she makes a big impression."[2]

The Rice style is clear: Tackle a subject, master it thoroughly and completely, perform at the highest level—preferably in front of movers and shakers—and then effortlessly become their protégée, accepting their mentoring on the path to power and success.

As she put it in an editorial for the Stanford campus newspaper, published in 1999 as she was retiring as university provost: "I hope each student takes the opportunity to learn something in-depth and to do something academically that he or she never thought possible. There is nothing as satisfying as getting to know one subject well. . . . Scholarly exchange is rigorous and demanding. It challenges those who participate to marshal facts and arguments in the face of intense critique. It sharpens one's intellect and, once mastered, enhances one's self-confidence. It is an investment worth making. . . . I hope every student pushes him or herself intellectually much farther than he or she ever imagined."[3]

If Hillary believes that the best path to greatness is through politics, elections, debating, advertising, attacking, rhetoric, and maneuver, Rice's career suggests that she has put her stock in excellent performance instead.

Hillary wants to be recognized by big-money donors, the national media, the political establishment, and, ultimately, the voters themselves in her quest for power. Rice has always banked on her ability to win admiration from important people to propel herself upward.

Hillary campaigns; Rice auditions. Hillary specializes in the pursuit of power, Rice in the pursuit of excellence. Hillary seeks supporters; Rice advances by attracting mentors. Hillary is always telling people—and having her friends and Bill tell them—just how good she is. Rice believes in letting people notice it for themselves. Hillary needs to win in competition with somebody else. Hers is a zero sum game: In order for her to win, she needs another person to lose. But Rice competes on her own, measuring herself against the standard of perfection. And when she falls short, as she did in trying to become a world-class concert pianist or an Olympic figure skater, she understands her limitations, accepts them, and moves on to seek excellence in another field.

Hillary has needed only one mentor in her life: Bill Clinton. But Condi has had a series of mentors: Czech refugee Professor Josef Korbel (Madeleine Albright's father), National Security Advisor Brent Scowcroft, Secretary of State George Schultz, and two presidents named Bush.

Hillary latched on to her husband-mentor and ran his campaigns, helping him become president; now she seeks what she sees as her just reward for the decades of service to his cause. Condi often met her mentors after they had achieved their success and attracted their attention not for how she could help them advance, but because she could enhance their performance in the office that they had already attained. (The exception is President George W. Bush, whom Rice tutored on foreign policy throughout his campaign and for whom she campaigned vigorously.)

If Hillary has worked day and night to achieve greatness, Rice has truly had it thrust upon her, usually by men in positions of authority and power who are dazzled by her performance. It is they who seek to advance Condi, not she who demands it.

Compare their college experiences: At the University of Denver, Condoleezza Rice won every imaginable award. Graduating with a B.A. in political science, the nineteen-year-old prodigy was the most honored member of her graduating class. Admitted to the honor society Phi Beta Kappa, she won the Outstanding Senior Woman Award, which the university said was "the highest honor granted to the female member of the senior class whose personal scholarship, responsibilities, achievements, and contributions to the University throughout her University career deserve recognition."[4]

Hillary won no such major award. But she used canny timing and political smarts to achieve recognition at her graduation anyway.

Having been elected president of the Wellesley student government, she demanded that a student—herself—be permitted to address the graduates at the ceremony to protest the Vietnam War and societal values. The speech put her on the national map for the first time and was widely featured in the news media of the day as an intelligent voice of student activism.

Two approaches to greatness. Neither woman was born great. Hillary struggled through political action to achieve it, while Condi strove only for academic excellence and achievement. Her recognition—and promise of greatness—came only through the acts of others.

Rice has always had a way of attracting attention and approval with her talents. Her first performance came at the age of four, at what she describes as "a tea for the new teachers in the Birmingham public-school system."[5] Reportedly able to read musical notes before she mastered letters, she matured into a world-class concert pianist, a skill she retains to this day. At the entrance to her National Security Administration office, Lemann reports, there hung "a large color photograph of Rice, the national security advisor, standing onstage with Yo-Yo Ma, the cellist, their hands clasped, their arms hoisted in triumph. They are acknowledging the audience's applause after having played a Brahms sonata together at Constitution Hall, in Washington."[6] In June 2005, she played at the Kennedy Center, accompanying a young singer with pulmonary hypertension at a fund-raiser to benefit the Pulmonary Hypertension Association.[7]

Her foreign policy interests date from her days as an undergraduate student at the University of Denver, where she was enthralled by the teaching of Professor Josef Korbel, a refugee who survived both Nazi and Communist tyranny in his native Czechoslovakia. It was in her international relations class that Rice began, as she puts it, "to fall in love" with foreign affairs.[8] And it was Korbel who noticed how brilliantly his earnest young student performed in class and helped to propel her ahead in her new chosen field. Rice recalls that Korbel inspired her to become a professor, choosing academia over a career in law: "He was nothing but supportive and insistent, even pushy, about me going into this field."[9] Rice wasn't Korbel's only protégée: His daughter, Madeleine Albright, became the first woman secretary of state—Condi's predecessor—when she was appointed by President Clinton.

As Rice moved up in her new area of expertise, getting her masters from Notre Dame and a Ph.D. from the University of Denver, she once again attracted notice and support, winning a fellowship to study at Stanford's Center for International Security and Arms Control. The coveted spot came with a stipend of $30,000, an office, and access to all of the university's facilities. It was supposed to last one year—but, as Rice's biographer Antonia Felix notes, "a few months after she arrived, Condi made such a big impression at a talk she presented to the political science department that she was asked to join the faculty."[10] Another successful audition.

Rice entered the Washington world as the protégée of the first President Bush's national security advisor, Brent Scowcroft. Scowcroft had first noticed Rice in 1984, when he met her at a seminar at Stanford University; then a junior faculty member, Rice challenged the foreign policy expert on his views. "I thought," Scowcroft recalled later, "this is somebody I need to get to know. It's an intimidating subject. Here's this young girl, and she's not at all intimidated."[11]

Scowcroft began arranging for Rice to be invited to seminars and conferences; by 1989, when his national security advisor appointment came, he appointed her to the National Security Council as the chief expert on the Soviet Union.

During her time in Washington, Rice became close to George and Barbara Bush. She came to work closely with the president, adding him to her list of supporters and fans. And when she served in these years as a member of Stanford's presidential-search committee, once more her work attracted attention—in this case, from the man chosen as the university's new chief, Gerhard Casper. He said he was "greatly impressed by her academic values, her intellectual range, her eloquence. . . . I have come to admire her judgment and persuasiveness."[12]

When she worked in Washington, her performance on the National Security Council attracted yet another admirer: former secretary of state George Shultz. In 1998, he introduced her to George W. Bush during a speech the president's son gave at a Republican fund-raiser in San Francisco. After the speech, the former secretary invited Condi to his home, where he was hosting a meeting of Stanford foreign policy experts with Bush. "Condi had a lot to say," Shultz told Lemann for his *New Yorker* article. As Lemann notes, Shultz listened: Condi had passed the audition. In 1999, Rice was invited to lead the younger Bush's foreign-policy team, a group referred to cheekily as the Vulcans. Once again, she passed the audition.

"The governor and Condi hit it off immediately," biographer Antonia Felix reports. "They were both exercise nuts—she would brief him on foreign policy while they huffed and puffed away on adjacent treadmills." Both were sports fans. "The two had a chemistry that created a bond of friendship, loyalty, and respect."[13]

When Bush invested in Rice—making her first his national security advisor and then his secretary of state—he was not only making a smart hire; he was giving her the grandest international stage to date. And in the five years since he came into office, her performance there has generated significant notice. In a sense, she is auditioning for the job of president before the nation's eyes and doing so with the same grace, skill, and success that has apparently marked her other performances. As America watches her navigate the dangerous waters of international diplomacy and witnesses her determination in fighting terrorism and fostering democracy at home and abroad, her command of her office is apparent. It would hardly be a surprise if America should reach out and ask her to run for president; such a movement would only follow the long line of people who have been impressed by this remarkable woman.

At this point, Rice contends that she is not running for president—any more than she was for a spot on the National Security Council when she attracted the notice of Scowcroft or for the national security advisor's job when she began her work with George W. Bush. Her traditional means of moving ahead is not to seek the office but to let the office seek her, moving ahead by ability, not ambition.

So Condi sits among us—without overt ambition and perhaps with some trepidation at the prospect of advancement—waiting to be asked.

Does she want to be president? She answered that question when she was eight years old and her parents took her to see the White House in Washington. "One day, I'll be in that house," she told her father.[14] And one doubts she had a staff position in mind.

Does she have the experience? Can she win? For all of Rice's ability to audition, perform, and win admiration, mentors, and advancement, she always seems to be underestimated. Or, in the word President Bush coined to describe himself, "misunderestimated."

As she was graduating from high school, she had her encounter with the soft side of racism—low expectations—when her guidance counselor advised her to get a job after high school because she was not "college material."[15]

Brent Scowcroft has remarked on Rice's "quiet demeanor."[16] But, as he notes, anyone who "thought they could push her around learned you could only try that once. She's tough as nails." Her Stanford colleague Coit Blacker put it more graphically: "The roadside is littered with the bodies of those who have underestimated Condi."[17]

Former CIA chief Robert Gates remembers how Rice accosted a Treasury Department official who tried to undermine her authority. "With a smile on her face, she sliced and diced him," Gates says. "He was a walking dead man after that."[18]

After her selection as provost of Stanford, the *Los Angeles Times* quotes one of Rice's best friends on campus as sensing "a certain condescension [toward Rice] in some of her meetings with senior deans or senior members of the faculty." But "she took 'em down a peg, took some very senior people down a peg, and that didn't sit well with a lot of them."[19]

The *New Yorker* reports that in her early years as Stanford's provost, she "gave the impression of an engaging rube, someone to be ever so gently patronized. A former Stanford administrator [said] that people thought of her as somebody who had real potential—to become, maybe, after a little polishing up in Palo Alto, president of a historically black college."[20] Janne Nolan, a fellow at Stanford who worked with Rice and is still a close friend observed: "I've watched it over and over again—the sequential underestimation of Condi. It just gets worse and worse. She's always thought of as underqualified and in over her head, and she always kicks everyone's butt."

Can Rice be nominated by the Republicans? Can she win election as president? The conventional wisdom says no. But are they misunderestimating her?

Rice would be more than just a black and female candidate. She would be Condoleezza Rice. To fathom how well she would do on the national political scene, one has only to explore who she is and how she has performed during her journey to the pinnacle of power.

WHO IS CONDI?

Condoleezza Rice's public record at the White House is of relatively recent vintage. It is her life story, more than her public career, that tells us why she could be a great president. Rice's biography is a unique story that bears elaboration. It is the story of an American black woman whose ability has carried her past once-lethal barriers with seeming ease and grace.

Condoleezza Rice has been defying odds since she was born in an all-black community in Birmingham, Alabama. Her family was solidly middle class, but in the Birmingham of those days, racial barriers could not be bypassed—even by money.

To grasp the atmosphere in segregated Birmingham is to enter a world beyond what most of us have experienced. When Condi's father went to register to vote in 1952, the registrar showed him a large jar of beans.[21] If he could guess exactly the number of beans in the jar, he was told he would be allowed to register.

At the age of four, Condi's parents took her to see the local Santa at Christmas—the first white man Rice can ever recall seeing. In this industrial Alabama city, racism lurked, literally, around the corner. Shopping as a young girl with her mother at a local department store, an employee told her she could not use the "whites only" dressing room and had to try on her clothing in a back storage closet. When Condi's mother refused and threatened to leave, the embarrassed employee relented. "I remember," Condi relates, "the woman standing there guarding the [dressing room] door worried to death that she was going to lose her job."[22]

On another shopping trip, a white saleswoman snapped at seven-year-old Condi: "Get your hands off that hat!" Rebuking the woman for her tone, Condi's mother encouraged her daughter to touch every hat in the store.

But the event that seared its way most powerfully into Rice's memory was the 1963 bombing of the Sixteenth Street Baptist Church, a few miles from her house. She heard the blast. Rice recalls the terror she felt, as an eight-year-old, that day. "These terrible events burned into my consciousness,"[23] she remembers. And, as America shook its head in disbelief at the murder of four girls in the blast, Condi Rice was mourning the two she knew personally—including Denise McNair, her kindergarten classmate, who was killed in the blast. "I remember more than anything the coffins, the small coffins, and the sense that Birmingham was not a very safe place."

Armed with a shotgun, her father joined the other men of the black community in night patrols to keep the Ku Klux Klan out of their neighborhood. Rice's biographer, Antonia Felix, recounts how "when a firebomb landed in Rice's neighborhood—a dud that didn't go off—John Rice [Condi's father] took it to the police and requested an investigation, but they would not conduct an inquiry."[24] It was in the crucible of that experience that Condoleezza developed her opposition to gun control and came to value what she sees as the Second Amendment guarantee of the "right to bear arms."

And on that blessed day when President Johnson signed the 1964 Civil Rights Act into law, banning racial discrimination in public facilities, the Rices went to celebrate by dining in a whites-only restaurant—their own personal civil rights demonstration. "The people there stopped eating for a few minutes,"[25] Condi recalls, as they entered the once-segregated facility.

Racism also followed her to the University of Denver, where her professor lectured the 250 students in his class on the genetic inferiority of African Americans, citing the pseudoscientific work of William Shockley. Rice simmered as her professor recounted Shockley's belief that "art, literature, technology, linguistics—all the treasures of Western civilization—are the products of the superior white intellect." "Rather than crouch down in her seat to avoid the onslaught," Felix reports, Rice "sprang out of her chair and defended herself: 'I'm the one who speaks French! I'm the one who plays Beethoven. I'm better at your culture than you are. This can be taught!' "[26]

The ever-present threat of racism followed Rice as she ascended the career ladder. On occasion, she struck back. While shopping with a friend at Stanford, Condi caught a whiff of it from a clerk who was showing her costume jewelry. Hostile remarks flew, and Condi let him have it: "Let's get one thing straight. You're behind the counter because you have to work for six dollars an hour. I'm on this side asking to see the good jewelry because I make considerably more."[27]

Yet Condoleezza's childhood is not just a saga of race and rage; it's also one of a middle-class young woman striving for excellence. As the *New Yorker* pointed out, "She was an only child, born to older (for that time and place: both Rices were over thirty when she was born), well-established parents, with a large supporting cast of relatives in addition to the community itself, and a long-standing family tradition of ambition and education."[28]

For all her childhood encounters with racism, Condi's life in Birmingham was one of relative privilege. According to Lemann, her parents "brought a special intensity" to her upbringing; she had "flute lessons and ballet lessons and French lessons and violin lessons and skating lessons and skipp[ed] two grades in school,"[29] entering college when she was fifteen. The name Condoleezza is an adaptation of the Italian "with sweetness"—an indication of the life of music and joy her parents had set out for her.

Condi's mother and father tried to shield her from the arrows of racism. As Rice told *Ebony* magazine, "Our parents really did have us convinced that [even though I] couldn't have a hamburger at Woolworth's, [I] could be president of the United States."[30] She was taught to "blast through the barriers."[31] She adds: "What's the alternative? Decrying the barriers? I tend to think that societies move largely through the force of individuals breaking barriers."

Rice told the 2000 Republican National Convention the story of "how her paternal grandfather had paid for tuition at Stillman College by selling cotton, and by encouraging her to apply for scholarships for Presbyterian ministers. "Granddaddy Rice said, 'That's just what I had in mind,'" she recounted. "And my family has been Presbyterian and college-educated ever since."[32]

Colin Powell—whose wife, Alma, was well-placed in the Birmingham black social circle in which the Rices moved—describes Condi's parents as "very accomplished professional people who expected nothing less from their children."[33] How did Condi and her family deal with segregation? "That's just the way it is," she remembers them feeling, "and don't let it be a problem. . . . We will go to extreme lengths to make sure that you get the same kind of cultural influences in your life that other children get."

The Rices remained surprisingly aloof from the civil rights protests that swirled around them. "My father was not a march-in-the-street preacher,"[34] Condi recalls. The Birmingham street demonstrations prominently featured children. When they were met with harsh police tactics, the resulting scenes of dogs snarling at young girls brought home to all of America how corrupting segregation was. But Condi's father "saw no reason to put children at risk," she recalls. "He certainly would never put his own child at risk."

Yet Condoleezza's father was an extraordinary man. As Antonia Felix relates it, "To say that John Rice was a tireless youth leader and educator is an understatement."[35] Rice was a minister, a teacher, a counselor, and a coach. He helped open the first Head Start center in Birmingham and helped minority kids find summer jobs. He got his masters degree in 1969.[36]

Rice explains that "my parents were very strategic. I was going to be so well prepared and I was going to do all the things that were revered in white society so well that I would be armored somehow against racism. I would be able to confront white society on its own terms."[37]

Rice's childhood was very different from Hillary's. At the Rodham household, there was no such stress on disciplined self-improvement, and, obviously, no sense of great obstacles to overcome. Hillary's own reports suggest that her childhood involved little of the structured nurturing and strict goal-setting that Rice saw. In *Living History,* Hillary notes gratefully that "I was lucky to have parents who never tried to mold me into any category or career. They simply encouraged me to excel and be happy."[38]

Hillary writes that she grew up "in a cautious, conformist era in American history"[39] and says her high school days resembled the movie *Grease* or the television show *Happy Days*.[40] She seems to have grown up without any of the structure or the demands that surrounded Condi. Where Rice's girlhood was a constant search for excellence, whether in skating, piano, or French, Hillary's appears to have been much more playful. Hillary describes how her family would gather around the television set and watch Ed Sullivan on Sundays.[41] She speaks of playing in the girls' softball league, ice-skating on the Des Planes River, and riding her bike "everywhere."[42] She notes with glee that she and her friend were "allowed to go to the Pickwick Theater by ourselves on Saturday afternoons," afterward repairing "to a restaurant for a Coke and fries."[43] In the future senator's childhood, there were no French lessons, flute lessons, or piano lessons, just a childhood of hanging out. Yet she too developed a sense of discipline and achievement.

Without parents constantly telling them what good people they are, many young people seek outside for affirmation of their evolving identities. Condoleezza Rice appears to have found her reinforcement at home before she went out into the world.

On the other hand, praise was rare in the Rodham household. So Hillary had to go outside her home to seek approval. She found it in movement politics at Wellesley and Yale and now finds it among her liberal friends. Her dependence on her peers for reinforcement explains a lot about her loyalty to the left. Nurtured in the "movement" as a sort of alternate family, Hillary seems instinctively to look to her ideological soul mates for support and succor.

Rice doubtless had frivolous times in her childhood, too. But overall her life seems to have been driven by infinitely more serious goals. From her college days at the University of Denver to her Phi Beta Kappa graduation at age nineteen and her later degrees at Notre Dame and Denver, it was clear that Rice had a special ability. After coming to Stanford to teach political science, she exhorted her students to live the same way. "Find your passion," she told them. "You've got four years in college, and if at the end of it you know what makes you want to get up in the morning, that's all you need."[44]

The one theme that *is* present in Hillary's youth, but absent in Rice's, is politics. Even the few pages Mrs. Clinton devotes to her childhood in *Living History* are filled with references to her youthful electoral triumphs. "I was elected cocaptain of the safety patrol,"[45] at elementary school, she tells us proudly. Her high school principal "asked me to be on the Cultural Values Committee."[46] She remembers that she "ran for student government president against several boys and lost."[47] Her consolation? She was "elected president of the local fan club for Fabian, a teen idol."[48]

And at Wellesley, Hillary was "elected president of our college's Young Republicans during [her] freshman year,"[49] before she saw the light and switched parties. And, of course, Hillary won the presidency of the college student government.[50]

Rice, on the other hand, was entirely focused on individual self-improvement. She never ran for any office in school and remained separate and apart, a prodigy who mastered every manner of musical instrument.

Hillary's adolescence was marked by constant social activism. From her precocious efforts—at thirteen—to discredit Chicago mayor Richard Daley's corrupt ballot count to rig the 1960 election for Kennedy, to her service as a Goldwater Girl in 1964, to her involvement in the McCarthy antiwar crusade of 1968, Hillary was a joiner and an activist.

But Rice, even as a black girl in the segregated Birmingham of the 1950s and 1960s, was more of a loner. As Alma Powell described the Rices: "They were not the generation that would get social change. They did not participate in sit-ins and marches. They were leery. In conversations with older people, you'd hear things like 'Oh, I don't know what's going to happen.' But there was no opposition to the movement—none of that."

As one of Rice's friends put it: "We don't all have a deprivation narrative." The Rice family did not need a handout or a hand up. Condi would move ahead on her own.

In the *New Yorker,* Lemann speaks of "the great intellectual divide of twentieth century black America—between W. E. B. DuBois, the radical proponent of political change, and Booker T. Washington, the advocate of self-improvement and not confronting the Jim Crow system." The Rices, he concludes, "were more on the Washington side."[51] Hillary came of age in the context of a movement—the antiwar student activism of the 1960s. Her memoir makes it clear that her political life really began at Wellesley, where she demonstrated on campus, defended the Black Panthers, and even traveled to Oakland, California, to work in the law offices of former communist Robert Treuhaft, one of the Panthers' lawyers. From the start of her adulthood, Hillary saw herself as an agent of social change, an activist in a political world, always part of a group, a phalanx committed to rearranging the world.

Hillary has always affiliated with her peers to achieve power. Once it was the antiwar movement; now it is the Democratic Party. In each case, she has sought her power as part of a community, an aggregation of like-minded individuals of whom she sees herself as spokesperson. In Rice's life, it has always been the individual who counted more than the group. Her dedication to the notion of personal excellence and self-improvement reflects her own coming of age as a parentally driven overachiever who needed to rise above the racism that kept blacks down in Birmingham.

In their differing backgrounds—and the life choices that flowed from them—Hillary and Condi reflect the different priorities of their political parties and the approach they take to the problems of social betterment, upward mobility, and race relations.

Like Hillary, the Democratic Party and its surrogate bodies deal with groups, seeking to enhance their cohesion and feeling of commonality. In industrial relations, labor unions emphasize the importance of collective action in bargaining, contracts, and strikes. The Teachers Union, for example, resists any attempt to pay teachers based on their individual performance—merit pay—because it fears that differentiation among its members will undermine unity and create divergent interests. The American Association of Retired Persons (AARP) rejects President Bush's proposal for private investment of a portion of the social security tax, worried that a large portion of the retirement community will be able to live off their investments and not need social security, weakening the political case for its preservation by diminishing its constituency. All focus is on the group, the aggregate. The message for members is clear: We all must hang together and move up or down as a unit.

In race relations as well, the Democratic Party emphasizes the group. Civil rights organizations and leaders stress the shared concerns of all African Americans, cutting across economic and class lines. They tend to focus on issues, like police misconduct that stimulate group identification. Again the message comes through: You are black; therefore, you are a Democrat. We all are.

So much stronger is the Left's sense of group and issue identification that feminist groups have largely opposed Republican women candidates who do not follow the feminist line on abortion rights, endorsing Democratic, pro-choice men instead. Indeed, in some cases, they have endorsed a pro-choice male Democrat over a Republican woman with identical views because of their loyalty to the Democratic Party. And innumerable issue groups—gay, environmental, Jewish, consumer, and handicapped—all seek to emphasize their group cohesiveness and common interests, walking a path that leads directly to the Democratic Party.

Individual upward mobility is alien to the credo of the Democratic Party. Those blacks who make it and leave their voting bloc are portrayed by some as traitors to the cause. Those who stray—from Clarence Thomas to talk show host Armstrong Williams to Colin Powell and even to Rice herself—are seen as apostates, shattering the unity that the Democratic Party has convinced African Americans they need.

The very logic of individual upward mobility flies in the face of the ideology of the group leaders who undergird the Democratic Party. If you move out of poverty on your own, what is the purpose of having civil rights leaders? If you gain pay raises because of your ability, why do you need a union? If a woman climbs the corporate ladder and no glass ceiling has kept her down, what is there left for a woman's group to achieve?

This environment is tailor-made for Hillary Clinton, who learned to speak, act, and think in a group. She is a pack animal, at her best when she is the spokesperson for others, especially when attacking the group's enemies. This is true far more for Hillary than for her husband: Where Bill has never felt limited by group identification—running for president as a New Democrat, triangulating to adopt Republican policies as president—Hillary tends not to challenge the party phalanx around her.

Nothing, of course, could be more alien from Condoleezza Rice's perspective or background. She came of age rejecting group identification and insisting on her ability, as an individual, to rise above the limits her race and sex imposed on her. In the course of her startling path to the top, she seems to belie the need for group cohesion or ethnic group advocacy.

And, in this spirit, she identifies most profoundly with the core belief of the Republican Party: That it is the individual who matters, regardless of circumstance, geography, race, sex, or even poverty. Her example proves that those with ability can move ahead on their own, without tilting the playing field or giving them preferential treatment. For those without the education or ability to move ahead, the Republican Party emphasizes the tools of individual self-improvement, like education, job training, personal investments, and the like—precisely the elements of Rice's own life and rise.

If the Democrats see individual upward mobility as a danger to group cohesion, the Republicans see the tendency to herd into a group and stick together as stimulating a sense of victimhood and class identification that is alien to true democracy.

Democrats accuse Republicans of callousness, saying they neglect those at the bottom and work only for the few who are well equipped to compete in life. Republicans accuse Democrats of wanting to enhance bloc voting by trying to keep the poor and minorities in a group, dependent on handouts from the political system for their upward mobility.

Each argument, of course, contains elements of truth. Republicans are not callous, but they are often self-involved and unfamiliar with what it means to be without any advantage. The Democrats' interest in group cohesion is not solely political, but the political rewards of identity politics have not escaped their notice.

A race between Condi and Hillary would put these competing philosophies on prominent display, since each candidate so completely mirrors in her own life the theme of her party. This could prove an advantage to Rice, whose public record of individual achievement is being played out in the headlines every day. Hillary's public record, on the other hand, is cloaked in ambiguity. Her tendency to speak of what "we" have accomplished blurs any real sense of her own personal role in serving the public good.

The popular impression, fostered by the media, is that Hillary Clinton has been a "good" senator, a moderate who has worked hard for her constituents. But what does that mean? What has she actually achieved?

The fact is she has been almost unrelievedly liberal. And she has been almost totally ineffective.

5

Hillary's Senate Record: The Grand Deception

David Remnick, writing in the *New Yorker* on July 4, 2005, reflected conventional wisdom when he referred to Hillary Clinton's "solid freshman term in the Senate."[1] Somehow the Clinton operation has spread widely the perception that Hillary has been an effective United States senator.

Effective?

According to the Library of Congress, Hillary has had a total of twenty bills passed since she entered the Senate. Of those, fifteen have been purely symbolic in nature. Only five Hillary Clinton bills of any substance at all have passed—two about 9/11, which any New York senator would have introduced and the Senate enacted, and three others.

Some record!

Here is a list of the bills Senator Hillary Clinton has passed in five years in the Senate.

Symbolic[2]

- Establish Kate Mullany National Historic Site
- Support the goals and ideals of Better Hearing and Speech Month
- Honor John J. Downing, Brian Fahey, and Harry Ford, firefighters who lost their lives on duty
- Recognize the Ellis Island Medal of Honor
- Name courthouse after Thurgood Marshall
- Name courthouse after James L. Watson
- Name post office after John A. O'Shea
- Designate August 7, 2003, as National Purple Heart Recognition Day
- Support the goals and ideals of National Purple Heart Recognition Day
- Honor the life and legacy of Alexander Hamilton on the bicentennial of his death
- Congratulate the Syracuse University Orange Men's Lacrosse Team on winning the championship
- Congratulate the Le Moyne College Dolphins Men's Lacrosse Team on winning the championship
- Establish the 225th Anniversary of the American Revolution Commemorative Program
- Name post office after Sergeant Riayan A. Tejeda
- Honor Shirley Chisholm for her service to the nation and express condolences on her death

Substantive

- Extend period of unemployment assistance to victims of 9/11
- Pay for city projects in response to 9/11
- Assist landmine victims in other countries
- Assist family caregivers in accessing affordable respite care
- Designate part of the National Forest System in Puerto Rico as protected in the Wilderness Preservation System

In the face of Hillary Clinton's reputation as an effective U.S. senator, this record of paltry accomplishment is sobering. As much as Alexander Hamilton, Harriet Tubman, and the American Revolution deserve our recognition, one thinks the voters of New York may have expected more of their junior senator.

Of course, Mrs. Clinton has been more effective in cosponsoring legislation that has been ultimately passed into law. But as anyone with even a nodding acquaintance with the legislative process can attest, cosponsorship means little. Senators routinely send around a form to their colleagues soliciting cosponsors on any of their important pieces of legislation. To cosponsor a bill just involves signing one's name and, occasionally, attending a press conference. To compensate for her failure to pass much of anything herself, she often grabs onto the legislative initiatives of others—usually prominent Republicans—for a free ride as a cosponsor.

South Carolina's Republican senator Lindsey Graham—one of the very House prosecutors who tried to throw her husband out of office—was amazed one day as he was about to hold a news conference to "drum up support for his bill to offer full-time benefits to military reservists."[3] Just twenty minutes before it began, Hillary's staff called and told him she wanted to join him as a cosponsor and appear at his press conference. Within several minutes, Hillary showed up. Graham recalls, "It seemed like a tornado came through . . . cameras started clicking like crazy because it was me and her." Hillary, of course, had nothing to do with the bill. It wasn't her idea. It wasn't her bill. But she made it her press conference, even if it meant—especially if it meant—pressing the flesh with a member of the "vast right wing conspiracy."

And Graham isn't the only GOP senator who's been visited by that tornado. She teamed up with South Dakota's senator John Thune, who sent Democratic majority leader Tom Daschle to an early political grave, to push for federal child support for members of the armed forces who die in the line of duty.[4] She jumped on the bandwagon with Kansas Republican senator Sam Brownback to urge Albania to have a clean election. She got together with Oregon Republican Gordon Smith to extend benefits to elderly and disabled refugees and with Rhode Island Republican Lincoln Chafee to reauthorize grants to technical institutions dealing with water resources.

But sometimes Hillary does come through in an important way. Republican senator Mike DeWine (R-Ohio) recounts how she battled successfully alongside him for the "pediatric rule" requiring drug companies to use scientific testing to determine if their medications are safe for children and at what doses they should be administered. "When she was first lady, Hillary pressured the FDA to approve the rule," DeWine said. "But the courts threw it out saying Congress had to authorize the FDA to issue the rule first. So when she got to the Senate, she worked hard with [Senators] Chris Dodd (D-Conn.), Ted Kennedy (D-Mass.), and me to pass a bill doing just that."[5] DeWine, a card-carrying Republican, credits Hillary with "holding the left in check" so that they would not load the bill with requirements as it passed the Senate, which would have doomed it in the more partisan and conservative House of Representatives.

Of course introducing legislation is only part of a senator's job. She also has a vote through which she can make her mark. As a Democrat, Hillary has proven to be nothing more than a reliable rubber stamp for the party leadership. According to the semi-official tally in *Congressional Quarterly,* Hillary has supported the position of her party more than 90 percent of the time in her Senate votes.[6]

Analyzing the key votes of each year, *CQ* found that Hillary supported her party position 97 percent of the time in 2004, 93 percent in 2003, 98 percent in 2002, and 96 percent in 2001.

Gilbert and Sullivan might have been describing Hillary Clinton's Senate record when they wrote:

> I grew so rich that I was sent[7]
> By a pocket borough into Parliament
> I always voted at my Party's call
> And I never thought of thinking for myself at all
> I thought so little, they rewarded me
> By making me the ruler of the Queen's Navee . . .

If only Hillary's ambitions were limited to running the Navy!

Mrs. Clinton has been a similarly knee-jerk supporter of the positions of the Democratic Party's interest groups. The AFL-CIO says she voted in line with their views 100 percent of the time in 2001 and 2004 and backed their opinions 92 percent of the time in 2002 and 85 percent in 2003.[8]

Her enslavement to the positions advocated by the big labor unions was most evident in June 2005, when she voted with the bulk of her party against the Central American Free Trade Agreement (CAFTA), the best hope for ending poverty in our beleaguered neighbors to the south. While the vote smacked of hypocrisy for many Democratic senators, it was particularly so for Mrs. Clinton, whose husband had staked his administration's prestige on pushing the North American Free Trade Agreement (NAFTA) through Congress in the opening years of his term. Hillary also voted against giving the president the authority to submit trade agreements with other nations to the Congress for fast-track approval—prompt consideration with no amendments or filibusters allowed. Bill Clinton pleaded with Congress annually, and in vain, for just such authority.

The liberal bellwether group the Americans for Democratic Action (ADA) similarly applauds Hillary's voting record, giving her a 95 percent score for each of her first four years in the Senate.[9] And Hillary's record, according to the ADA, was markedly more liberal than the Democratic Party's as a whole. While all Democrats put together backed the ADA position 85 percent of the time, Hillary supported them on 95 percent of the votes.[10]

Hillary Clinton's votes all echo the liberal line in the Senate.[11]

- She opposed the ban on partial birth abortions and came down against criminalizing harm to a fetus during an attack on the mother.
- She voted against the repeal of the 1995 tax on social security benefits.

- She opposed limits on class action lawsuits and was against capping medical malpractice damages against obstetricians and gynecologists.
- She opposed a travel ban to Cuba.
- She voted against oil drilling in the Arctic and opposed Bush's changes in the Clean Air Act.
- Hillary opposed a constitutional amendment banning gay marriage.
- She backed extending the ban on assault rifles for ten years.
- Hillary opposed President Bush's prescription drug benefit plan.
- She was against Bush's tax cuts and opposed repeal of the estate tax. At every chance, she voted to cut the amount of the tax reductions Bush proposed.

The only exceptions to her party-line voting were her support for the Iraq War and her votes for appropriations to fund it, her uniform support for tough antiterrorist measures, and—in an attempt to curry favor with the media—her opposition to nullification of the FCC rules making media consolidation easier.

In short, there is nothing in her voting record that a preprogrammed machine could not have done as well. In neither her record of passed legislation nor her votes is there the slightest reflection of the effective senator she promised New Yorkers she would be.

In fact, even her failed legislative proposals reflect the minimalist, bite-size strategy of the Clinton administration. On health care reform, her proposals have been very limited. While she has introduced thirty bills on the subject, most just nibble around the edges of the problem; none goes to the heart of the difficulties Americans face getting health care. None of them has anything to say about drug prices or coverage for those without insurance (except for tax credits to persuade small businesses to offer coverage). None of her bills calls for any reforms in the managed care system—no patient bill of rights, nothing concerning appeals of HMO decisions, or even any legislation about the right to sue to protect one's health care.

To bolster her record on health care legislation, Hillary cites the passage of the Nurse Retention Act as an achievement. She says she worked with Republican senator Gordon Smith of Oregon on its passage, but an interview with a member of the Senate staff who followed the bill closely provides a different picture.

"The bill was going to pass anyway," he said. "Everybody backed it. There was no heavy lifting involved. Lots of senators filed essentially the same bill and neither Smith nor Clinton deserves much credit for its passage. To say that Hillary was instrumental in passing it is pure puff."[12]

In the areas of education, children, and families—other fields on which she said she would focus—her record is similarly devoid of substance. Her website's leading claim is that she "supported the 'No Child Left Behind' education bill,"[13] but President Bush's legislation is hardly Hillary's achievement. She says she "sponsored an amendment to promote teacher and principal recruitment," but doesn't say that it failed. And she boasts that she released a statewide report on New York's schools, an accomplishment more worthy of a civic group than of a United State senator who supposedly has the power to get things done.

Indeed, the only area in which Hillary has amassed a good legislative record is on fighting terrorism. She has pushed hard for threat assessments on bioterrorism, to protect the food supply, promote benefits to children of terror victims, increase homeland security grants, investigate how to secure radioactive materials, require annual inspections of high-risk sites, identify potential terror sites, encourage bomb-scanning technology, and improve protection at our embassies.

But none of these bills has passed.

Despite Hillary's voluble pledge to fight for Israel in the Senate as she represented the state with the largest Jewish population, not a single piece of legislation, resolution, amendment, or even any expression of the sense of the Senate in the entire period of 2001–2004 even mentioned the name "Israel."

And yet Israel went through perilous times during Senator Clinton's first four years in office. Almost one thousand Israelis were killed and countless more seriously wounded and injured in suicide or homicide bombings while Hillary sat in the Senate. Were the United States to lose a proportionate number of their citizens, that would mean more than a hundred thousand dead—thirty-three times the loss on 9/11. And yet not a single bill from Hillary expressed support for Israel as it faced a most trying hour.

To review Hillary Clinton's legislative proposals—most of which have not passed—is also to grasp what a big spender she still is. Hillary's record is a far cry from the fiscal conservative she pretends to be as she wags her finger at the Bush deficit and demands financial restraint. In fact, as the National Taxpayers Union noted, she has "topped the Senate by sponsoring or cosponsoring 174 spending bills."[14]

While none of the individual causes she has voted to fund is irresponsible in itself, if most or all them had passed, it would have added vast sums of red ink to the budget deficit. Specifically, Hillary has proposed additional spending to improve military housing, study the impact of defense base closures, keep open facilities on closed defense bases, upgrade armed forces medical readiness, increase aid to Kosovo, promote new communications technology, renovate aircraft hangars, preserve benefits for the remarried spouses of dead servicemen, increase aid to blind veterans, and expand VA programs . . .

. . . and to expand the Teacher Corps, raise school standards, upgrade foster care, study the health impact of dilapidated schools, increase education aid to New York, expand adult education programs, recruit more principals for schools, renovate more school buildings, raise grants for higher education, increase vocational education, expand VISTA, and study long-term child development . . .

. . . and to coordinate the environmental health network, increase FHA hospital grants, expand eligibility for children's health programs, spend more on nurse retention, establish a national health-tracking network, expand grants to research eating disorders, provide more money for respite care, give a tax credit for health insurance costs, analyze developmental disability data, expand lead paint removal, provide more services on the 211 phone number, give more health care information to patients, fund insurance for employees at atomic weapon plants, pay for mental health treatment for the elderly, underwrite female anti-AIDS programs, fund training for long-term care ombudsmen, study drug effectiveness, increase rape convictions, and diminish minority health disparities . . .

. . . and to increase funding for emergency conservation programs, protect the Finger Lakes region of New York, and expand the Puerto Rican national forest system . . .

. . . and to improve voter security systems, expand the electricity grid, give a tax credit for bonds that promote economic development, expand the work opportunity tax credit, increase funds for regional skills alliances, increase business incubator programs, expand unemployment insurance, and increase manufacturing extension aid.

Notice a theme in the verbs which presage all of her legislation? "Expand," "upgrade," "increase," "renovate," "fund," "establish," "pay for," "underwrite," "provide," "study," "create," "aid," "give a tax credit for"—all tax-and-spend verbs.

On their own, most of these bills are good ideas. Though Senator Clinton is on record as opposing most of the Bush tax cuts, she has yet to propose a single bill to cut federal spending in other areas or to augment federal revenues to pay for these programs. Like a person on a shopping spree, Hillary seems to view each new item on her wish list as essential. It's only the total tab that seems out of whack.

Has Hillary Clinton delivered for New York outside of the Senate chamber? Has she fulfilled her pledge to bring jobs to upstate New York?

Her accomplishments on this front are pitifully limited. She claims to have "facilitated" a $93 million contract between the Postal Service and Lockheed Martin Systems in Owego, New York.[15] She also lists grants of another $180 million for such projects as the Rochester fast ferry, improvements at Fort Drum, work on Buffalo's

Inner Harbor, a projected "Rooftop Highway" across northern New York, aid for the Nassau Medical Center, and other, unspecified projects.

Even taking her claim for credit for these accomplishments at face value, generating $300 million of aid or contracts for New York is pathetic for a state with a $600 billion economy. Can she truly claim that these achievements are sufficient to underpin a presidential candidacy?

In her description of her work in the 108th Congress (2003–2004) on economic development, she seems to confuse her role as senator with that of a conference organizer. She cites how she "helped bring people together all over the state, including the meeting between Broome County businesses and defense contractors in Binghamton; the Farm-to-Fork meeting in Syracuse; the Microcredit launch in Albany; and most recently, the Broadband conference in Delhi."[16] She boasts of cosponsoring the Twenty-First-Century Nanotechnology Research and Development Act, which did become law.[17] She gives herself credit for dozens of small grants for defense procurement in New York State, but their total comes to $160 million—hardly a significant part of the more than $400 billion in the national defense budget.[18]

And when New York City bid to host the Olympic Games in 2012, Hillary accompanied Mayor Michael Bloomberg to Shanghai to press the city's case. But, as with so many of her initiatives, it turned out only to be a photo opportunity. New York lost out to London, finishing last among the four finalist cities.

In the realm of foreign affairs, she says she "visited Israel" in 2002, "supported" legislation to help victims sue state sponsors of terrorism, "condemned" suicide bombings, "called on President Bush" to pressure Arafat to end terror, "delivered a statement on anti-Semitism," and "called on Secretary Powell to raise the issue."[19] As these verbs suggest, Hillary has accomplished nothing of substance in the realm of foreign policy. No legislation. No achievements. Nothing but talk.

In a sense, Hillary fits in well with the modern U.S. Senate. The days of the titans are over. This is not the Senate of Daniel Webster, Henry Clay, Richard Russell, Lyndon Johnson, or Hubert Humphrey. Senators cast more than six hundred votes each year, but the vast majority are along party lines, with very few defections. And the issues tend to be inconsequential. All the important decisions are made off the floor and even away from the committees, in broad bargaining between the leaders of the two parties and the White House.

The floor votes often concern obscure points, procedural motions, or feckless attempts by the minority to disrupt the majority's cohesion on an issue by offering poison pill amendments. An automaton could do as well as the average senator—or congressman—in casting votes on the floor.

The only real horse-trading among the individual members involves pork-barrel projects for the home folks. Specific appropriations for particular projects in their districts absorb a huge portion of the average senator's time. Most questions of high public policy are resolved far away from the floor, in conference committees and informal phone consultations between legislators and executive branch staff.

The overall point is clear: Hillary Clinton is rated highly by her colleagues, her opponents, and New York State voters not so much for what she *is,* but rather for what she *is not.* It's only in contrast to her history of partisanship and shrill dogmatism that Hillary looks good to them. She is running against herself—and, for the moment, she seems to be winning.

THE 9/11 DECEPTION

But Hillary's favorable image among her New York constituents doesn't really have much to do with her legislative record. It largely stems from the perception that she did a lot to help their state in the aftermath of the attacks on 9/11. This assault, coming only nine months after she took her seat in the Senate, has come to define, more than anything else, her job performance in the eyes of her constituents. It was during this period, as the city was rebuilding, that this woman from Illinois, Arkansas, and Washington seemed to pay her dues to become a real New Yorker.

The attacks of 9/11 not only cost us more than three thousand lives; they also decimated lower Manhattan, the core of New York's financial center. A vast proportion of its office space vaporized on that dreadful day. The effort to rebuild became the focal point of all New York political activity. All other issues—job creation, education, mass transit, crime—fell to the side as all eyes turned to Washington to learn what help New York could expect from the federal government.

New York's relationship with Washington has always been tenuous. At the height of the city's fiscal crisis, which brought it to the verge of bankruptcy in the mid-1970s, President Gerald Ford declined to intervene and help keep New York City afloat financially. The *New York Daily News* captured the mood of its readers with its famous headline: "Ford to City—Drop Dead."[20]

At the time, it was the intervention of New York State that saved the day for its largest city. After 9/11, however, only Washington could get the city back to work. It was in these months that Hillary Clinton made her reputation. But her efforts were exaggerated by her publicity machine and by her own public statements, which sometimes descended to outright fabrication.

One week after the attack, Senator Clinton was introduced by Katie Couric and interviewed by Jane Pauley for a segment on the *Today* show. Perhaps Hillary wanted to be more than a spectator to the horrific events in her newly adopted state; perhaps she wanted to appear to be a victim of the tragedy and draw sympathy from the national audience. But for whatever inexplicable reason, Hillary went on the air and convincingly—but falsely—suggested that Chelsea had narrowly missed being on the grounds of the World Trade Center at the very moment when the planes hit the towers. Even Chelsea herself later contadicted her mother's perverse lie.

A voiceover that introduced the segment told the viewers, "At that moment, she was not just a senator, but a concerned parent." Hillary then went on to say that Chelsea had "gone on what she thought would be a great jog. . . . She was going to go around the Towers. She went to get a cup of coffee—and that's when the plane hit. . . . She did hear it, she did."[21]

Apparently, Chelsea was unaware that she was supposed to have been in imminent danger at the World Trade Center on 9/11. In an article she wrote for *Talk* magazine two months later, in November 2001, Chelsea earnestly told the story of her experiences on 9/11. According to Chelsea, she was staying at a friend's apartment in New York's Union Square area, about *three miles* from Ground Zero, when her hostess called her from work to tell her about the attack. Chelsea wrote that she "stared senselessly at the television" as she saw the terrorist plane strike the Towers. She mentioned no jog, no narrow escape, no coffee shop. She was obviously nowhere near the Trade Center and heard nothing but the television broadcast. The jogging incident? Hillary simply made it up.

To this day, the *Today* show has never corrected the story or even challenged Hillary about the sickening lie on their show to their national audience.

And one more thing: Hillary never repeated this dramatic and *false* story in her own memoirs. She did address Chelsea and 9/11 in her book and thanked her Secret Service agent, Steve Frischette, for his "calm presence" in reaching Chelsea in "lower Manhattan." No mention of Chelsea's brush against danger, no mention of the 9/11 lie.

This episode raises serious questions about a presidential candidate. What kind of person would make such a claim? What kind of person would craft such a story to exploit this tragedy for political gain, instead of merely being grateful that her daughter had been out of harm's way?

But Hillary's deception about her role in the aftermath of 9/11 neither began nor ended with her *Today* appearance.

The reality of what Hillary really did falls far, far short of the image she has propagated. In his recent book *After: How America Confronted the September 12 Era,* Steven Brill, a well-known and respected journalist and the founder of Court TV and *American Lawyer* magazine, tells the story of what really happened as New York went—hat in hand—to Washington for the federal aid it needed to restore its downtown business district. As Brill's painstaking account shows, it was New York's other senator, Chuck Schumer, who did most of the heavy lifting.

Brill details the hour-by-hour process of piecing together support from Congress and the White House for aid to New York City to offset the devastation of the 9/11 attacks. His account is filled with the details of Schumer's efforts; Hillary has very, very little to do with it. (The bias is not Brill's: He has a reputation as a political progressive who was supportive of Clinton during the impeachment imbroglio.)

According to Brill, Schumer got busy early on Wednesday, September 12—the morning after. "Schumer spent Wednesday morning deploying his staff," he recounts. Some staff members were dispatched to assess the demand side of the equation and "canvassed city and state agencies" to evaluate their needs. Others worked the supply side of the street, checking out what other members of Congress thought could be made available. By the following day, Schumer had heard the figure $20 billion bandied about Washington as the cost of reconstruction and safety—*for the entire nation.* As Brill reports, Schumer reasoned: "If the whole country needs $20 billion, well then, so too did New York."[22]

At this point, Brill mentions Hillary's role for the first time. Schumer sold the $20 billion figure to Senate majority leader Tom Daschle (D-South Dakota), while Hillary convinced the Appropriations Committee chairman, Robert Byrd (D-West Virginia).

Later on Thursday, Brill reports, the two New York senators met with President Bush in the Oval Office to press for aid to help in the recovery from the attacks on the Trade Center. (They were accompanied by their colleagues from Virginia, representing the Pentagon.) On the way into the meeting, Brill reports, "Clinton told Schumer he should do the asking."

Brill recounts an emotional scene in which Schumer told the president that he had heard it said that America needed $20 billion to recover from the tragedy. "Well, Mr. President," Senator Schumer continued, "New York needs $20 billion, too." He recounted the extensive damage the attacks had inflicted on the city's infrastructure, transportation system, and inventory of office space.

Brill says that the president seemed to hesitate at first, but then stepped up to the plate: "New York needs $20 billion?" he asked. "You got it," he decided.

Beyond her silent presence, Hillary seems to have played no real role in the process.

But then the $20 billion commitment seemed to come unstuck. Schumer, a former congressman, heard that the House Appropriations Committee was going to earmark the sum for recovery without specifying that it was to go to New York. Schumer went nuts. Brill reports that he screamed at Bush's chief of staff, Andrew Card, implying that he would embarrass Bush during his visit the next day to Ground Zero if he did not keep his commitment. The $20 billion was earmarked for New York.

Where was Hillary? Brill's authoritative account heaps all the credit on Schumer's shoulders. None for Hillary.

But the Brill/Clinton saga does not end here.

When Hillary learned that Brill was writing a book about the aftermath of 9/11—and giving Schumer most of the credit—Hillary invited the journalist in for a chat. Cornering him at the ceremony commemorating the first anniversary of the attacks, she told the author: "I hear you want to talk to me about your book."[23]

Brill and Clinton had several discussions, including several in which she was off the record. In one on-the-record interview, however, she insisted that "the person responsible for aid to New York City was not [Senator] Chuck Schumer but Hillary Clinton."[24]

Concerned that Brill would downplay their boss's role, Hillary's staffers worked hard to persuade him of Clinton's active work with the victims of 9/11, providing what Brill said was "an elaborate story, with an elaborate subtext of memos and phone calls—a long, long story."[25]

But when Brill checked out the facts, he noted that "None of it turned out to be true. . . . They gave me documents and phone calls and things like that which just plain never happened."[26]

Brill told WABC Radio's Steve Malzberg that he was astonished at Clinton's fabrications, which were "in the cause, actually, of shooting down Chuck Schumer getting a bunch of pages in a book." Brill ruminates that "it sort of takes your breath away when you think about it."[27]

Brill, who interviewed many 9/11 families, learned in his reporting that Hillary took great pains to limit her availability to the relatives of the victims. Brill says: "I think [Senator Clinton] has begun every statement she's ever made in her life, about the families of the victims, by saying she's met innumerable hours with the families of the victims."[28] But the real stories Brill unearthed were quite different.

Brill gives the example of one family, the Cartiers, who lost a member of their family on 9/11. "This family had tried repeatedly to get Hillary Clinton to meet with them," Brill said. "And always the staff said: 'Write up a memo. We don't meet with any families unless they write to us first and tell us what they want to meet about.'" Brill reports that the Cartiers told him—"and these are not, you know, people who are political"—that the only time families would be invited to meet with Hillary would be "at a press conference."[29]

"Meanwhile," said Brill, "Senator Schumer took time out on a Sunday to meet with the Cartiers—no reporters, no cameras, no nothing."[30] Mayor Giuliani and Schumer also helped the Cartier family in their search for their relative's remains.

In sum, Brill said, "What stunned me is that one person would try to steal away the credit from the other person, especially when everything I was hearing from the families is that Schumer" had been the driving force behind the fund-raising effort.[31]

But Brill's story was not about just giving a few extra pages of coverage to Schumer. It represented a threat to the central myth upon which Hillary predicated her claim to the affections of her adopted state: that Hillary Clinton had delivered after 9/11.

Hillary's staff, of course, wasted no time in denouncing Brill's accusations. In a classically Clintonian rapid-response counteroffensive, Clinton spokesperson Philippe Reines told Bill O'Reilly on his top-rated Fox News program *The O'Reilly Factor* that "Brill's accusations are completely false" and called Brill's charges "an obvious last-ditch effort to jump-start anemic book sales." Reines couldn't resist twisting the knife one further turn: "It's hard to understand why Mr. Brill would choose to exploit such a horrible tragedy in this manner," he lamented.[32]

Brill retorted that he could prove what he said about Hillary denigrating Schumer and her staff giving him false information. "If Hillary Clinton will simply release me from my pledge to keep those conversations off-the-record, I will be delighted to tell everybody in chapter and verse exactly what she said to me."[33]

Interviewed by NewsMax, the conservative newswire service, Brill said that he had "contemporaneous notes" of an extended conversation he conducted with Hillary at Ground Zero a year after the attacks. Those notes, he contended, "will clearly show who is telling the truth. Anyone who wants to look at those notes or who wants to even hear what I told people who I was talking to in my office after I came back from my conversation with Hillary—that off-the-record conversation—it will be plain as day that she went out of her way to denigrate Chuck Schumer so that he wouldn't come off in my book as the person who was responsible for helping New York City."

And so it goes with Hillary Clinton: Unwilling to confront President Bush with even something as unquestionable as a request for aid to New York in the wake of 9/11, she has resorted instead to fabricating her record, embellishing it where it fell short and inventing a role she did not have in the response to America's greatest contemporary tragedy.

Aren't we entitled to ask for a little more integrity from our leaders?

6

Rice at the Pinnacle

Rice never needed to exaggerate her record or credentials. By the late 1980s, she was an expert on the Soviet Union. A Stanford professor and prolific author, she focused on the key elements of the U.S.-Soviet relationship: arms control and German reunification.

Learning to live with the Soviet Union was no easy task. Ever since President John F. Kennedy found himself nose-to-nose with Soviet leader Nikita Khrushchev during the Cuban Missile Crisis—with a nuclear holocaust hanging in the balance—the goal of superpower politics had been nuclear and conventional disarmament. With twelve thousand warheads on each side aimed at the other, the threat of global annihilation hung heavily over international diplomacy.

For her part, Rice had become steeped in Russian culture and life, even mastering the difficult language. But the main thrust of her expertise related to the military issues that dominated the American-Soviet relationship. Rice's first book, *The Soviet Union and the Czechoslovak Army 1948–1963 Uncertain Allegiance,* focused on the military relationship between Moscow and the satellite government in Prague. As a fellow at the Stanford Center for International Security and Arms Control, Rice studied the details of the SALT and START Treaties between the superpowers to reduce, bilaterally, the size of their nuclear arsenals.

As an academic, Rice also focused on the hundreds of thousands of Soviet and Western troops that faced each other in Central Europe—seemingly separated by only a tripwire, in an uneasy equilibrium that threatened at any moment to erupt into a world war. At the core of this issue was the separation of Germany into its Western Democratic and Eastern Communist parts, originally the zones of occupation assigned to the victorious allies after Hitler's defeat. This division was demarked by the Berlin Wall, which had become the symbol of the Cold War.

Rice's emphasis on disarmament and German reunification never seemed more relevant than in 1989, when she accepted Scowcroft's invitation to serve as the National Security Council's expert on the Soviet Union. Mikhail Gorbachev had come to power in Moscow in the mid-1980s and, through a policy of *perestroika* (restructuring) and *glasnost* (opening), hoped to revive the fortunes of the Soviet Union. His policies promised an end to the era of political repression and heavy-handed bureaucratic economic regulation that had stopped Soviet progress.

When British prime minister Margaret Thatcher decided that Gorbachev was "someone we can do business with," changes that had seemed unattainable before suddenly became within reach. Perhaps the arms race could stop, and German reunification could begin after all.

Rice's mandate was clear: to apply her academic knowledge and perspective to the rapidly unfolding developments at hand. As former President Bush recalled in his memoir, *A World Transformed:* "I had chosen Condi because she had extensive knowledge of Soviet history and politics, great objective balance in evaluating what was going on, and a penetrating mind with an affinity for strategy and conceptualization. . . . She was charming and affable, but could be tough as nails when the situation required."[1]

Rice's focus throughout the first Bush administration was on balancing the superpower relationship, reuniting Germany, reducing armaments, and managing the transition out of the Cold War. While these issues were doubtless important, some have expressed a belief that she missed the underlying changes taking place. The thirty-four-year-old Condoleezza Rice of the late 1980s, they feel, showed too much loyalty to the group thinking of the era and too firm a tether to academic theory about the Soviet Union—a view that was being made obsolete even as she was explaining it. They have even charged her as largely responsible for the first President Bush's laggardly approach to a post-Soviet world.

But, in her very limitations during the term of Bush the elder, it's possible to see the beginnings of the evolution that came over her by the time of George W. Bush's presidency. An intellectual academic, steeped in the learning of the past, Rice began to grow into a moral leader determined to bring on a very different future. The story of this growth—and of her willingness to admit her mistakes along the way—contrasts vividly with Hillary's great difficulty in ever admitting errors or mistakes, and her consequent inability to grow.

It is important to understand the limitations of the Condi Rice of 1989–1991 to appreciate fully the growth that was manifest in the period after 2000.

For decades, those who shaped America's international policies have faced a basic question: Should our foreign policy be dominated by the need to advance our national interests, or should we give a priority to promoting moral values and democracy around the world?

The dichotomy between the two approaches is well explored in the opening two chapters of Henry Kissinger's book *Diplomacy,* where he compares the administrations of presidents Theodore Roosevelt (1901–1909) and Woodrow Wilson (1913–1921).[2] Kissinger contrasts Roosevelt's emphasis on projecting American power and protecting our economic, diplomatic, and financial interests with Wilson's idealistic commitment to the global rule of law and the formation of an international community of nations. Roosevelt's "dollar diplomacy" stressed the need to back up American companies and investors with our military prowess and sanctioned frequent armed intervention in the Western Hemisphere to support our interests when they were threatened. By contrast, Wilson called on the United States to support global values of peace and freedom, emphasizing national self-determination, freedom of the seas, and the rights of neutrals. Even when Wilson led America into World War I, he did so by calling for a "peace without victory."

The story of Condoleezza Rice's foreign policy journey is really the tale of a migration from a national-interest perspective, informed by her academic work, to a focus on morals and value issues in the aftermath of 9/11.

Scowcroft, Rice's boss in the first Bush presidency, had come of age on Kissinger's staff during the Nixon and Ford administrations, where the balance of power, geo-political correlation of forces, and national interest dominated. After his election in 1980, Ronald Reagan swept aside the value-neutral calculus that dominated during the Kissinger years and called for an aggressive war on Soviet totalitarianism, describing Russia as our adversary and labeling the Moscow regime as the heart of an "evil empire." Scowcroft was relegated to a more minor role as the focus of foreign policy shifted from national interest and superpower balance to values and an anticommunist moral crusade.

But Scowcroft, with his new employee Condi in tow, came back to the White House under the administration of Bush the elder, as a focus on national interests in foreign policy came back in style. President Bush had, after all, cut his diplomatic teeth during the Nixon/Kissinger years as U.S. ambassador to the United Nations and in the quintessentially *realpolitik* position of director of the Central Intelligence Agency.

As an academic, Rice was devoted to the concept of "realism"—as propounded by author and global philosopher Hans Morgenthau—which emphasized the pursuit of national self-interest in international affairs, a focus that was often expressed in terms of military power. Ann Reilly Dowd, writing in *George* magazine, said that Rice "came to see the Cold War, not as a war of ideas between communism and democracy, but something more primordial—a raw contest between two great competing powers with conflicting national interests."[3]

And, as the first President Bush's Soviet expert, Condoleezza Rice was right in the middle of that superpower relationship. She prepared the president for four summit meetings with Gorbachev and organized a "seminar" of Russian scholars who briefed Bush on the Soviet Union in February 1989. She traveled with the president to Poland and to Germany to mark the fall of the Berlin Wall. She also traveled with Bush to his first meeting with Gorbachev in December 1989.

At their first summit, off the Mediterranean island of Malta, Bush introduced Condi to the Soviet leader. "This is Condoleezza Rice," Bush told Gorbachev. "She tells me everything I know about the Soviet Union." Gorbachev looked her over, startled in that setting by this slender, thirty-five-year-old African American woman.

"I hope you know a lot," the Russian replied.[4]

(Condi's venture into the world of superpower summits was not entirely auspicious. The meeting was set to take place on two ships, the American *Belknap* and the Soviet *Maxim Gorky,* floating near each other in the normally calm Mediterranean Sea. As the summit opened, though, a storm convulsed the region with gale-force winds, stoking sixteen-foot waves that violently rocked the ships. The storm was so violent that what reporters came to call the "seasick summit" had to be postponed for a full day while the world's superpower leaders waited for nature to calm down. Rice, who is no expert swimmer, had to negotiate the treacherous passage between ships in a small launch; her prominent role in the international summit proved less intimidating than those waves. "The most frightening moments," she said, ". . . were when you had to . . . walk down this plank, with ocean on both sides, winds howling, rain coming right at you, and since I'm not the world's best swimmer, I thought, 'Well, I guess I could die here for my country and no one would ever know it.' ")[5]

At the start of Rice's tenure in Washington, Gorbachev seemed like a blast of fresh air. She recalls how "Gorbachev said, 'We want the United States to stay in Europe too. The United States is a European power.' And given the history, where we had always believed, and where all of us had been taught . . . that it was the principal goal of Soviet power to get America out of Europe, this was an extraordinary statement, and it stuck with everybody." Rice was particularly impressed when "It became clear to me that [Gorbachev] had in mind the Soviet Union as a legitimate actor in Europe, not feared in Europe but respected in Europe."[6] But change was moving too fast for Gorbachev—and for Bush and Condi, too. As Rice recalled of those years, "Events were unfolding so quickly that you would make a policy or make a decision or arrange a meeting, and before you could get there, everything had changed, and indeed the world changed. . . ."[7] The Soviet leader was learning that democratic reforms would bring not unity, but more dissent, to his dominion. With the newly freed voices of glasnost demanding an end to the monopoly of the Communist Party on political power, Gorbachev found himself more and more a communist and less and less a reformer.

Voices from the Right in the United States—among them defense secretary Dick Cheney and his undersecretary Paul Wolfowitz—were calling for Bush to break with Gorbachev and back democracy in Russia. After all, reformer or not, Gorbachev was still a communist, committed to the continuation of one-party rule in the Soviet Union. The "neo-conservatives," echoing the views of former president Ronald Reagan, felt that the only real way to guarantee American security was to transform the world into an assemblage of democracies. They didn't want to get along with the Soviet Union so much as change it.

But Rice, President Bush, and Secretary of State James Baker were reluctant to get too far ahead of the reforms in the Soviet Union, lest they undermine Gorbachev and turn Moscow back over the hardliners. Their emphasis was to avoid providing any pretext for the old guard to come back—and, at all costs, to avoid embarrassing Gorbachev. As Rice put it, "When you have so much power, you have to be careful not to get in the way of historical events that are going your way. Too heavy a hand by the U.S. might have provoked a counterreaction."[8]

But the neo-conservatives got the essential point that Rice, Bush, and Baker were missing: that the Soviet Union was falling apart. As Nicholas Lemann wrote in the *New Yorker,* "Academic friends who saw [Rice] in those days would say that what mattered most was the palpable, pervasive rot in the Soviet system, not that Gorbachev was a great leader; Rice, in the time-honored manner of the academic turned government official, would say, 'If you could see the intelligence I see, you'd understand that I'm right.' "[9]

The decay that was infecting the Soviet Union began not only to force democratic changes inside Russia, but also to create a centrifugal force that drove away the Soviet Eastern European satellites and eventually dismembered the Soviet Union itself.

An early result of this decomposition of the once formidable Soviet empire was the demise of the Brezhnev Doctrine, which gave Moscow the right of armed intervention in any Warsaw Pact nation that strayed from socialism into the forbidden ground of democracy. Gorbachev effectively discontinued the doctrine, announcing unilateral reductions in Soviet troop presence in Eastern Europe. While Bush, Baker, and Rice welcomed the cut, they missed the larger point—that without Soviet troops in Eastern Europe, the satellites were free to throw off their chains. Years later, Rice admitted that her focus on the quantitative issue of Soviet troop reductions caused her to "miss . . . the revocation of the Brezhnev Doctrine."[10]

Once Rice grasped what was really going on, she became a strong supporter of aid to the newly emerging anticommunist movements in the satellite countries. When Solidarity, the Polish labor union led by Lech Walesa that pushed for an end to Soviet domination, was legalized—after years of repression under martial law—Rice helped convince Bush to provide all possible economic aid to Poland to bolster the forces of freedom.[11] And, in July 1989, Bush visited Hungary and Poland—with Condi at his side—to speak up for democratic movements in both countries and to hold out the implicit promise of American support.

Still, many conservative critics felt that the Bush administration was missing the point. Jacob Heilbrunn wrote in the *New Republic* that "the Bush administration seemed almost sorry to see the Soviet Union go. Its instincts were to hold the Kremlin's realm together so as not to upset the existing geopolitical order."[12] Worried about what would happen to the Soviet nuclear arsenal, they almost appeared to want to keep the old bipolar world alive for a few more years.

To some, Bush, Baker, and Rice had begun to act a bit like the British POW Colonel Nicholson, played by Alec Guinness in the film *Bridge Over the River Kwai.* Assigned by his captors to build a bridge, he developed such momentum and pride in his work that he found himself struggling to complete the bridge even as the Allied army bore down on the Japanese, bringing with them the promise of his liberation.

The long-standing agenda of the Cold War—strategic arms control, conventional military cuts, and German reunification—seemed larger to them than the more important context of the total dissolution of our prime adversary.

Rice saw the liberation of Eastern Europe primarily in the context of German reunification. After all, it was in East Germany that the Soviet Union had legal standing, stemming from its role in the World War II victory. As Rice described it in a 1997 interview, "Germany unification was perhaps the most important issue of this entire period . . . because that is where the Cold War began, and that was the only place that the Cold War could end. . . . I don't think that there is anything else that demanded the attention in the administration that German unification did."[13]

John Prados, a Rice critic, wrote in the "Bulletin of Atomic Scientists" in 2004 that "Rice saw the problem of ending the Cold War as one of securing the reunification of Germany under the Western umbrella without triggering a response from Moscow."[14] Gorbachev cannot have foreseen that the centrifugal forces that spun off Eastern Europe would then dismantle the Soviet Union itself. In the end, all of the "People's Republics" that comprised the Soviet Union left Mother Russia and formed independent countries—nations that are becoming more and more pro-Western as this is written.

As the Soviet Union teetered on the brink of dissolution, Rice recalled that she felt as if she were "in a race to try to finish the business of ending the Cold War with Gorbachev still in power. It was a very delicate balance, a very short window of opportunity, because the Soviet Union had to be strong enough to sign away its four-power rights and responsibilities, but not strong enough to stop it. . . . I always try to remind people that some year and a half, fifteen months after we managed to unify Germany, the Soviet Union broke apart, so the timing couldn't have been better."[15]

Why did Rice place so much emphasis on the legalisms of the situation? Why was she acting like a deathbed lawyer, getting the Soviet Union to sign away its estate before it expired? Rice had her reasons: If the Soviet Union failed to sign away rights to the part of Germany its troops had won in World War II, she feared, lingering Soviet resentment could one day undermine the peace of Europe. She said it was important to "be respectful of Soviet interests, while indeed watching a set of events that were very quickly unraveling Soviet power in Eastern Europe."[16]

Rice makes the point that the outcome of the events of 1991 need not have been so positive. "Was it inevitable," Rice asked, "that Germany unified on completely Western terms in NATO; that Soviet troops went home with dignity and without incident; that American troops stayed; that all of Eastern Europe was liberated and joined the Western bloc? No, it was not inevitable—and that leaves a lot of room for statecraft."[17]

But it was not only Eastern Europe that was leaving the Soviet sphere of influence; it was the non-Russian states that comprised the Union of Soviet Socialist Republics—particularly the Ukraine. Incredibly, Bush chose to try to prop up the USSR rather than stand by and watch our principal adversary self-destruct. In August 1991, he traveled to Kiev in the Ukraine and gave what his critics have derisively called his "Chicken Kiev" speech, urging Ukrainians to stay loyal to Moscow and condemning "suicidal nationalism." Bush declared, "We will maintain the strongest possible relationship with the Soviet government of President Gorbachev."[18]

Some critics have been quite harsh about Rice's performance during this difficult passage. Laura Flanders, author of *Bushwomen: Tales of a Cynical Species,* writes that "the foreign policy staff were split, and most of the men who worked with her then and . . . again today have good reason to remember Rice as the 'expert' who was doggedly, disastrously wrong on the most important development in her area of expertise."[19]

As circumstances evolved, Russian Federation president Boris Yeltsin, who had quit the Politburo and the Communist Party, began to challenge Gorbachev's policies. The Soviet leader, still committed to Marxist-Leninist ideology (or whatever version of it they were actually practicing by then), refused to dilute the power of the Communist Party in response to Yeltsin's urgings. The other Soviet Republics, particularly the Ukraine, became increasingly restive under Gorbachev's rule.

But even more ominous was the threat of a restoration of hard-line communist rule as the old Stalinists in Moscow began to resent Gorbachev's reforms. The fear of the old communists reasserting themselves in Moscow led Bush and Rice to focus on propping up Gorbachev and ignoring the growth of Yeltsin and the democracy movement.

Condi didn't much like Boris: "He struck me as mercurial and difficult," she said. (Could she have meant drunk? Bill Clinton told me that it was not until his fourth summit with Yeltsin that he ever saw him sober.) As Nicholas Lemann has observed, "if you're dealing with Condoleezza Rice and you're messy and undisciplined, you've got two strikes against you right away."[20]

In 1989, Rice even had a physical face-off with Yeltsin during his visit to the White House. With Gorbachev still in power, President Bush did not want to bolster Yeltsin by meeting with him in the Oval Office, so it was arranged that Bush would drop by a meeting at the National Security Council's offices, where Yeltsin would be huddled with Rice and Scowcroft. Scowcroft arranged for Yeltsin to enter the White House through the basement, in hopes that a low-profile visit would minimize the offense to Gorbachev. Condi was chosen to shepherd the truculent Russian governor through the process.

But Yeltsin wasn't buying it. When his car arrived at the basement door, he protested. Sitting in his limo with arms folded, Boris refused to leave his car unless Rice promised to take him to the Oval Office. "This isn't the door you go in to see the president," he complained.[21] Told that he would be meeting with Scowcroft, Yeltsin snapped, "I've never heard of General Scowcroft." Rice and Yeltsin glared at each other in silence as the minutes ticked by. Neither would budge. Finally, Condi said, "You might as well go back to your hotel." Her firm stand paid off. Yeltsin relented and tamely followed the young woman to the National Security Council Office where President Bush soon appeared, much to the Russian's delight.

Eventually, the communist hardliners got fed up with Gorbachev's reforms, launching a coup d'etat in an attempt to restore the old-time religion—atheism—to the Kremlin. The coup attempt fizzled, but Yeltsin used the opportunity to oust Gorbachev and replace him with a democratically elected president—himself. The Soviet Republics seized the chance to break from Russia, and the current global map took shape.

What is most remarkable about the Condoleezza Rice of these years is not how shortsighted her critics feel she may have been during her first stint in Washington, but how much she had grown by her second. The Rice who served Bush I was inclined to see the world in terms of superpower relationships, correlation of military forces, spheres of influence, and the legal rights that emerged from World War II. The Rice who returned to serve Bush II, it appears, had been transformed by a stint at a most unlikely finishing school for prospective foreign policy leaders: a mid-career tenure as provost of Stanford University. It is Condi's management record at Stanford that gives us the best indication of what kind of president she might be.

IS RUNNING A UNIVERSITY LIKE RUNNING A COUNTRY?

When Rice became provost of Stanford University in 1993 at the age of thirty-eight, she took over responsibility for an institution of 14,000 students, 1,400 faculty members, and a $1.5 billion annual budget. The usual adjectives surrounded her appointment: She was the youngest, the first female, and the first African American provost in Stanford history. When she left the post in 1999, she was roundly hailed as one of the best in the university's history.

Rice was tested in every possible way during her six years on the job. In her time at Stanford, she confronted many of the same problems that face a president of the United States. She had, for example, to overcome a huge budget deficit largely by cutting expenditures—mostly administrative—offending many and stepping into the midst of controversy.

Most wrenchingly, this young, African American, female provost had to maintain academic standards in the face of insistent and incessant demands for racial and gender preferences in the granting of tenure to faculty members. At the same time, she had to resist the demands of angry Californians to do away with affirmative action altogether and close the portals of opportunity for so many who, like Condi, would make it on merit if only given a chance.

Provost Rice also had the responsibility of preserving Stanford's academic standards, particularly for undergraduates, and was forced to wrestle with a persistent housing shortage for students and faculty.

More than most governors—and certainly more than any senator—Rice's ability to cope with administrative, policy, and management problems was tested at Stanford. After all, running a university prepared Woodrow Wilson—and, to a lesser extent, Dwight Eisenhower—to become effective national chief executives.

The lessons Rice learned at Stanford are obvious in her description of the difficulties a manager faces: "As an executive," she says, "you're always asked to make important decisions about which your knowledge base is relatively slim. Someone might ask me to support a million dollar physics telescope. I don't know a lot about that, but I can ask hard questions and get a sense of whether it's important and prioritize it against other issues."[22]

Rice came to Stanford as provost as part of a house cleaning necessitated by the resignation of university president Donald Kennedy following a scandal over Stanford's use of federal grant money. According to the *Los Angeles Times,* "The school acknowledged billing the government for, among other things, depreciation on a school yacht and the cost of flowers, parties and furniture at Kennedy's campus home."[23] To replace Kennedy, the board chose Gerhard Casper from the University of Chicago. Rice first got to know Casper when she served on the search committee that appointed him. Casper was so impressed with Rice's ability that, only eight months after taking office, he turned the tables on her and named her provost of the university—the number two job on campus.

Of course race and gender—as well as ability—played their parts in her selection. The *Los Angeles Times* noted that "Stanford had a history of complaints regarding alleged bias. Many considered Rice's appointment an effort to address those concerns in dramatic fashion."[24] Casper himself told the *New Yorker,* "It would be disingenuous for me to say that the fact that she was a woman, the fact that she was black and the fact that she was young weren't in my mind."[25]

But more than race or gender, it was Rice's youth that appealed to Casper. He told *Stanford Magazine* in 1993 that "her age was clearly something of great importance to me. I want to bring the younger faculty into University governance. In a way, that is the single most important aspect of this appointment, beyond her outstanding personal qualifications and background."[26]

Casper let Rice assume tremendous responsibility, and the two fashioned a kind of collaborative relationship. Rice became what he described as more of a "deputy president" than a provost, with "a completely discrete set of responsibilities."[27]

Casper asked Rice to become "the crucial person in university budget matters," and "central in the process of academic appointments and promotions and, as such, is responsible for maintaining the highest academic standards for the university."[28] As the *Los Angeles Times* put it, the job "required grit, skill, political savvy and a sublime degree of self-confidence."[29]

As usual, Rice had been vaulted over barriers to become the provost. As Casper acknowledged, normally a candidate for provost "should have been dean . . . should at least have been a department chair." But he named Condi anyway because "I was absolutely convinced that she was competent."[30]

Rice has described the Stanford job as "the toughest" in her life, and tough it was.[31] Her first task was to wrestle with a $20 million annual budget deficit she had inherited.[32] Rice was met with incredulity when she promised to balance the university's budget in two years. In a desperate effort to eliminate its deficit, Stanford had already cut almost $40 million in the years before Condi's arrival; most bet that additional reductions would be almost impossible to find.[33]

"There was sort of conventional wisdom that said it [balancing the budget] couldn't be done, that [the deficit] was structural and we just have to live with it," said Coit Blacker, a fellow professor and friend of Rice's.[34] But "Condi said: 'No, we're going to balance the budget in two years.' It involved painful decisions but it worked, and communicated to funders that Stanford could balance its own books and had the effect of generating additional sources of income for the university."[35] Rice proved as good as her word: She managed to balance the budget in two years, slicing expenditures by $16.8 million and raising $3 million in new revenue. She slashed the budgets of academic departments and student services and faced the need for layoffs.

But Rice didn't see the task as unusual. "I actually don't think of this as a budgetary crisis," she said. "This is just managing in the 90's. Every American institution out there is going through the same questions."[36]

Stanford's current president, John Hennessy, described Rice's moves as painfully necessary and even courageous. "No one likes layoffs, especially universities, because there are so many interpersonal relationships," he has said. Rice's work on the budget "was enormous," he adds. "We could have had problems lingering for ten years, easily, if it wasn't addressed in dramatic fashion."[37]

At the time, Rice's moves were met with heated opposition—and made "more brutal," according to the *Los Angeles Times,* "by the imperious way she carried them out." Albert H. Hastorf, a former Stanford provost, says, "She was extremely autocratic in her style. She didn't brook anyone disagreeing with her."

Challenged to consult with a faculty committee in deciding which cuts to make, Rice said, "I don't do committees." She told the *Financial Times* in a 1995 interview, "I am direct. . . . Sometimes someone has to draw a line between informing, consulting and deciding."[38]

In retrospect, it's interesting to note how calm Rice was as she was remaking the university—and stabilizing its finances—through these cuts. She left no hint of indecision or even pain as she went about evaluating the problem methodically. If anything, she erred on the side of decisiveness and often failed to consult as much as she might have with others.

After she left Stanford, Condi reflected: "Maybe I was too much of a hard-ass. Maybe if I had it to do over, I'd be a little gentler."[39]

But Condi's hammer was coated in velvet. "She's very charming, very gracious, but she can really come down on you hard when she has to," her friend Kiron Skinner says. "I have seen her be very, very tough on people; they have to back down."[40] When a group of students held a hunger strike over Rice's decision to lay off Chicana dean Cecilia Burciaga, Condi's friend asked her if the demonstration bothered her. "I'm not hungry," she reportedly replied. "I'm not the one who's not eating."[41]

The Rice who made tough decisions about cutting Stanford's budget is not always a figure of great warmth. She was no touchy-feely administrator, anxious to mend fences and stay popular. She often gave the impression of insensitivity as she made her cuts. But, at the same time, she seemed to pass a message as she laid people off: It's not personal; it's just business.

This is yet another key contrast between Rice's style and Hillary's. Mrs. Clinton's confrontations are quite personal and usually tied with her demand for political fealty in everyone she works with. When she laid off the staff of the White House Travel Office, it was not to trim the budget as Rice did at Stanford, but because she feared their political disloyalty. Frequently paranoid, she sees enemies around every corner. Where Rice cut Stanford staff to balance the budget, one can see Hillary making cuts to eliminate opposition and quell disloyalty.

There was no hint of favoritism in Rice's management style at Stanford. When you are downsizing your staff, cutting jobs at every turn, any indication of helping old friends or punishing former enemies is likely to be quite apparent and widely condemned. But Rice's record is devoid of any such accusation. On the other hand, there is no sense that she abstained from cutting jobs to court key constituencies. She was all business—no personal favoritism, no politics.

By contrast, Hillary is always political. When she masterminded the selection of Bill's cabinet in 1992, she became expert at balancing the expectations of each constituency group in her and Bill's effort to design an administration that "looked like America." For Hillary Clinton, jobs added or subtracted are opportunities to enhance loyalty, reward friends, and purge enemies and threats. No decision is made without politics and personality injecting themselves into the equation.

But Condi worked differently, and the results were gratifying. "As the university's No. 2 administrator," the *Los Angeles Times* reported, "Rice is widely credited with helping the school regain its footing during the 1990s after red ink and a financial scandal threatened to engulf it."[42] Indeed, Rice's administration was so successful that during her term Stanford was able to finance the most intensive period of rebuilding and construction in the university's history, costing $1 billion.[43] In 1996, Rice was able to report that Stanford had not only eliminated its deficit, but had a $14.5 million budgetary reserve.

But balancing the budget—Rice's first challenge as she became provost—was not her biggest one. Next came an issue that went to the very heart of who she was and how she had gotten to the top: affirmative action.

Starting in 1995, Rice was under constant scrutiny and pressure to increase the number of tenured female and minority faculty. As tensions mounted, the provost—a black woman herself—was attacked for failing to use affirmative action criteria in awarding tenure to black and female faculty.

A 1998 Stanford report by the Faculty Women's Caucus, "Status of Women on the Stanford Faculty," contended that Stanford did not hire women in proportion to the number of available female doctorate holders. "By 1997," the report said, "a mood of crisis and low morale characterized not only many women faculty members but also many junior and minority faculty at Stanford. Several denials of tenure to women and minorities contributed to an uneasy sense that the university's commitment to diversity was declining and squeezing out many of the strongest women and minority junior faculty members."[44]

Slamming Rice, the faculty women said, "Stanford's recent policy towards the promotion of both women and minorities could be characterized as one of benign neglect. Once hired, all faculty are left to achieve on their own merits. But benign neglect—or color and gender 'blindness,' as it can be called—is often an inadequate response in the context of a culture and community largely shaped by a history of systemic inequality and exclusion."

But even as Rice faced criticism on her campus from the Left, the atmosphere in California as a whole had turned decisively against affirmative action programs of any sort. In 1996, the voters of the state overwhelmingly passed Proposition 209, which outlawed any use of gender or racial criteria in state contracting, admission to public universities, and government hiring.[45] Condemning affirmative action as a form of inverse racism, the Right wing demanded an end to all race or gender-based preferences.

Rather than bow to the pressures of the Right or the Left, Rice charted a clear, unambiguous centrist course and stuck to it through all manner of pressure and criticism. Admitting that she herself was a product of affirmative action, Rice endorsed using racial and gender preferences in admitting students and hiring faculty. "I am myself," she said, "a beneficiary of a Stanford strategy that took affirmative action seriously, that took a risk in taking a young Ph.D. from the University of Denver. The president of the university did, after all, appoint a thirty-eight-year-old black female professor provost who had never been a department chair."[46]

What had been good for her, she said, was good for others. "I support affirmative action in higher education," Rice said. "It makes the student body and the administration more integrated. It's accelerating the integration of all strata of society . . . [so] we don't have to wait one hundred years. I think it has to be done very very well."[47] But Condi cautioned against diminishing standards to accommodate the need for diversity: "I find those who are too concerned with minority students' particular experiences exhibit 'reverse racism.' It is equally bad to be patronized as to be disliked."

Yet, as much as she backed affirmative action in hiring faculty members, she strongly opposed it in granting tenure. Essentially, Rice's view was that the university should go out of its way to hire minorities and women and give them the tools to succeed, but then insist that they measure up, without regard for race or gender, before granting them a permanent place on the faculty.

Rice urged "an extremely broad, aggressive approach to the hiring of women and minorities at the junior level," but said that using racial or gender criteria in awarding tenure would compromise standards and the pursuit of excellence.[48] Calling the temptations of affirmative action "a real slippery slope," she consistently refused to give in to demands that she favor minority and women professors in granting tenure.[49]

In taking this position, Rice was echoing her own experience when she arrived at Stanford and joined the faculty. Back then, the chairman of the Political Science Department had told her, "We have a three-year period here and then you have to be renewed. And nobody's going to look at race. Nobody's going to look at gender; and you don't get any special breaks; and you surely don't get any special breaks when you come up for tenure."[50] Condi's reaction then, and later as provost, was the same: "Well, yeah, that seems perfectly fair." But the pressure from the Left mounted as a series of popular women and minority faculty failed to get tenure:

- In January 1997, assistant professor of anthropology Akhil Gupta from India was denied tenure (although Rice let an advisory board review the rejection, and as a result, he eventually was tenured).[51]
- In April 1997, the dean of the School of Humanities and Sciences denied tenure for popular assistant professor of history Karen Sawislak, even though her department had recommended that it be granted.[52]
- In December 1998, Linda Mabry, a popular black professor who taught international business law for six years, abruptly resigned,

citing the law school's "inhospitable environment" for minorities. Mabry said she was overlooked as a potential choice to head up a new law school program.[53]

- In February 1999, Robert Warrior, an assistant professor of English, was denied tenure by the dean of the School of Humanities and Sciences.[54]

While Rice refused to budge on the broad issue of using affirmative action in granting tenure, she dramatically improved the university's recruitment of women and minorities to the faculty.

In 1994, Rice set up the Faculty Incentive Fund, which distributed money to fund additional faculty positions where an outstanding woman or minority candidate was available who did not meet the criteria for any current openings on the staff. Under this initiative, taken despite university-wide cutbacks in spending, forty-six new minority or women faculty were hired at Stanford.[55] By the time Rice left her post in 1999, the number of women on Stanford's faculty had risen to 19 percent, almost double the 11 percent level of ten years earlier.[56] During the same period, minority faculty rose from 7 to 14 percent. There was even progress in granting tenure to women faculty: By the end of Rice's term as provost women were getting tenured 45 percent of the time, while only 38 percent of the male faculty were winning tenure.[57] The Stanford Law School had a woman dean, and the university had thirteen women department chairs, up from only two when Rice took over.[58]

Still, as she left, the *Stanford Daily* reported that there was "widespread discontent with the hiring and tenure rates of minorities and women" at the university.[59] So much so, in fact, that a group of current and former Stanford faculty brought a four-hundred-page complaint to the attention of the U.S. Department of Labor, alleging widespread discrimination against women in hiring, promotion, and tenure. The irony of this suit, triggered by the policies of a female provost, was not lost on the media. As the *Los Angeles Times* wrote, "Improbably, the youngest provost in Stanford history and the first black and woman to hold the post helped prompt a Labor Department probe into the treatment of women and minorities."[60] According to students for environmental action at Stanford, "it appears that Stanford University has yet to get it." One of the complainants agreed: "It will require this kind of external intervention to change the culture."[61]

Despite the ongoing criticism of Rice's actions from the Right and the Left, she walked a narrow line she felt was right, backing opportunity by supporting affirmative action in hiring of the faculty but insisting on standards by opposing the awarding of tenure based on racial or gender criteria. Rice rejected the path that would have been both politically correct and most conducive to peace on campus. As a black woman, she would have become very popular with her natural constituencies if she had adopted the pose of the woman who did more to integrate Stanford. Her resolute refusal to bend on diluting standards for tenure won her no friends and brought no peace. But she stuck to it and prevailed.

Had she caved-in on using affirmative action to grant tenure, there would have been no lawsuits. But Rice stood doggedly on principle, pleasing neither Left nor Right.

On the other hand, she could have made herself the poster girl for the Right wing by opposing all affirmative action. She could have followed in the footsteps of Samuel Ichiye Hayakawa, whose face-offs with rebellious minority students during his tenure as president of San Francisco State College won him sufficient statewide popularity to propel him to the U.S. Senate in 1976.

California in the 1990s was ripe for a black woman to come along and try to eliminate affirmative action completely. Had Rice taken such an extreme track, she would have acquired an instant political constituency. But she was intellectually true to her own past—to her debt to affirmative action—and she resisted the chance to take a demagogic posture against it.

Her tenacity in the face of protests, lawsuits, complaints to the federal government, campus demonstrations, and even hunger strikes showed a toughness and a tensile strength one would admire in a president—though it may also hint at a certain stubbornness that could hurt her in such a political role.

In the microcosm of the affirmative action question, we see the vast gap that separates these two women: Hillary and Condi. In both her policy and her style, Rice's handling of the controversy at Stanford differs sharply from how Bill and Hillary handled the same issues nationally.

Early in my time working with President Clinton at the White House, he flagged the danger he saw ahead in California's pending Proposition 209. Only a few months after the 1994 election, he and Hillary spoke to me about how they should handle this new hot-button issue. Should they side with those who wanted to end affirmative action, or remain loyal to the core constituencies of the Democratic Party?

At first, the president wanted to explore alternatives to affirmative action, examining ways to accomplish the same goal—helping minorities and women—without using gender or race preferences. He and I discussed modifying affirmative action to grant preferences to those in poverty, regardless of gender and color, and to give advantages to businesses located in the inner city and owned by neighborhood people without reference to race or gender. After all, half of the poor people in the United States are white, and many live in households headed by a white man.

But Hillary soon ended this flirtation with moderation. She saw great danger in disappointing the black and feminist groups that supported the Democratic Party and warned Bill that Reverend Jesse Jackson might oppose him for the 1996 Democratic presidential nomination if he waffled on affirmative action.

Hillary pointed out that many middle-class blacks and professional women felt they needed affirmative action to get ahead in their workplace or win government contracts. Diluting the program to give preference to poor people, regardless of race or gender, might strip them of their privileges—and, she argued, they are the core of the Democratic Party.

Eventually the Clinton administration opted to continue using race and gender preferences while pledging to avoid quotas or firings based on sex or color. But throughout the entire process of the Clintons' tortured deliberations, all focus was on the political impact of different possible positions on their chances for reelection. On the merits, President Clinton told me that he agreed that it would be better to base affirmative action on criteria other than race and gender. But Hillary convinced him that politics would not permit such a deviation from the party line.

(In her Arkansas years, Hillary was more willing to chart an independent course. She supported legislation that required teachers to take tests and specified that those who failed had to be fired. But by the time she hit Washington, this independence from the line laid down by the unions and other interest groups had long been purged from her thinking; now she backed racial and gender preferences even to the point of lowering admission criteria as a result.)

Rice didn't have to steer a path through this difficult terrain. She just didn't care about popularity. She took what she thought was the proper course, even though it antagonized both the Left and the Right.

During her tenure at Stanford, Rice also faced complaints that the university had subordinated its traditional teaching duties to its role as a research institution. Students said that they were too often shuffled off into large lecture halls and had little opportunity for personal interaction with their busy research-oriented professors. Here, Rice was more user-friendly and sympathetic to the students. She was determined to change the academic environment—to put teaching first—and insisted on more faculty-student interaction, smaller class sizes, and more seminars for undergraduate freshmen and sophomores. Most notably, she set up a Sophomore College where she and four other faculty personally taught fifty selected second-year undergrads who came to campus early for an intensive two-week academic program. Eager to increase the university's focus on language study, Rice raised writing requirements; she also revamped the science curriculum and reformed the student guidance system.

Behind these measures, one senses Rice's appreciation of the importance of the mentor-protégée relationships that have meant so much in her spectacular rise. The students seemed to realize how much of herself she was prepared to invest in their education. When she left the university, in a gesture that seems right out of *Goodbye Mr. Chips,* the student newspaper ran an editorial headlined: "Farewell Provost Rice: Condi leaves a legacy as a powerful administrator who cares about students." The editorial continued: "We've interacted with her as students in her sophomore seminar, benefiting from one of the many improvements in undergraduate education that she approved. Or we've heard her advice to pursue our passions."[62]

Her efforts to raise standards triggered such a surge in freshman applications that they hit an all-time high in 1999. Perhaps just as gratifying, Stanford also now had the highest percent in its history of freshmen who were invited to attend who then actually chose to enroll. Rice's successor, John Hennessy, complimented her efforts: "We've done more in a short period to change the first two years of college than any comparable institution."[63]

Hillary Clinton, of course, wrote a bestselling book, *It Takes a Village,* urging just such a mentoring approach to raising and educating the young. She wrote of the importance of finding role models for children so that they could become sober and responsible citizens. But when it comes to education—an issue particularly crucial to women voters—Condi has *lived* what Hillary has written about. After all, Condoleezza Rice is a professional educator. Hillary Clinton is not. Condi's record is one not just of advocacy, but also of action.

THE ROLE OF A NATIONAL SECURITY ADVISOR

When Rice initially went back to Stanford, she says, she wasn't longing to return to Washington. In 1995 she commented: "I don't suffer from Potomac fever in the way it affects many people who have worked in Washington and spend the rest of their lives wanting to go back. I can say in all honesty that I don't spend a waking moment thinking about whether to go. I had a chance to finish so much in those two years that I have no thirst to try to do it again."[64]

But soon Washington was calling—this time in the form of George W. Bush.

Rice's relationship with the Texas governor deepened throughout the late 1990s as he prepared for a presidential race. To describe her as an advisor would miss the point of the relationship. She seems to have served as a kind of catalyst to Bush's thinking, helping him assimilate, synthesize, and effect the advice he was being given. The intimacy of the Bush-Rice relationship reflects not the arm's-length dialogue one would expect of an advisor and a candidate—or president—but rather a shared journey, undertaken by the candidate in the company of his trusted expert, through the labyrinths of foreign policy.

Bush's friendship with Rice was strengthened by their shared love of sports—football in particular—and exercise. Bush had once owned a baseball team, and Rice grew up learning about football from her father. They shared a sense of humor, and both understood the need to keep what was said between them out of the newspapers. But it seems to have been Rice's spiritual heritage that did more than anything else to build the relationship. Their shared commitment to injecting spiritual concerns into their public policy work animated the relationship and made it closer.

Evan Thomas discussed the Bush-Rice relationship in *Newsweek:* "Superficially, Bush and Rice are opposites: the rich white boy from Texas who goofed off in school; the middle-class black girl who was a grind. But in fact they are well matched, and not just by a well-publicized mutual fondness for working out and watching sports on TV. The two are possessed of a certain defiant independence, almost an orneriness. They know what it's like to be underestimated, and they take obvious pleasure in going their own way. Deeply religious, the Presbyterian Rice and the Methodist Bush share a messianic streak. Rice's real job is to help steer Bush's black-and-white moral impulses in the murky, morally ambiguous real world. It is a tricky course, but in a sense, her whole life has prepared her for it."[65]

Or as Rice puts it, "Well, first, my faith is a part of everything that I do . . . and it's not something that I can set outside of anything that I do because it's so integral to who I am."[66]

President Bush is a person of instinct, gut, and impulse; his views on foreign policy stem, in large part, from his sense of right and wrong. Rice, by contrast, is a skilled academic, used to assessing evidence dispassionately and applying scientific method to public policy analysis. The two complement each other very well.

As Thomas writes, "During the campaign, Bush would sometimes blurt out a foreign-policy 'instinct,' and it would be up to Rice to make sense of it. This could take some doing."[67] But beyond the working relationship, Condi developed an intimacy with the entire Bush family. As the *New Yorker* noted, Rice not only works closely with Bush, but sees him during leisure hours, when other staff members do not—for example, watching football games.[68] In fact, "her primary off-hours companions seem to be George and Laura Bush."

The Bush campaign of 2000 was Rice's baptism of fire into the world of electoral politics. She addressed the Republican National Convention during prime time, when it was sure to be covered by national television, and served as the campaign spokesperson on foreign policy. No longer was Rice backstage. As the governor of Texas—a man with no foreign policy experience at all—tried to convince America that he could handle a world filled with crisis, she was presented as his expert witness.

After Bush was elected, he selected Rice to serve as his national security advisor—the second most powerful foreign affairs job in the government. In that role, Rice's relationship with Bush grew even closer, to the point where observers say she seemed to be on Bush's side of the membrane that separated him from his advisers.

Rice was always careful to keep her views understated, sharing them primarily in private with the president. But, because of her enormous influence, people were always wondering what she was thinking. *Newsweek* reported that "Rice is quiet, respectful, anonymous—but firm, just the way the president wanted it. Rice's aides call her the 'anti-Kissinger,' meaning that she does not need to show off her influence or present herself as a master global strategist. . . . That may be in part because Rice is not a strategic genius, but no one doubts her power. Rice's aides also refer to her (affectionately) as the 'Warrior Princess.' Rice has Bush's complete confidence; she speaks for the president, and everyone knows it. The harder question is how much she influences his thinking and his decisions."[69]

Rice told *Newsweek* that "her job as national security advisor is to sharpen arguments, not squelch them or flatten them."[70] As Bush's then-treasury secretary, Paul O'Neill, told Evan Thomas: "She drives towards clarity. Then he [the president] decides what the consensus is."

In a sense, the story of the Bush-Rice relationship during her tenure as national security advisor is one of a shared journey, in which the president and his counsel evolved a view of international affairs grounded in their shared moral view and sense of mission. The chain of influence between the expert and the president seems to flow both ways. Condi brings to Bush her academic grounding, the perspective of history, and a dose of *realpolitik*. Bush contributes his unwavering grasp of good and evil, his values-oriented approach to international issues, and his ability to integrate his spiritual bearings with his conduct on the global stage.

As Nicholas Lemann described the Bush-Rice relationship in the *New Yorker,* "Rather than her simply guiding him through the unfamiliar world abroad, it looks as if something more complicated and interesting were going on: he's actually influencing her, and she seems to be performing for him the immensely useful service of transforming shorthand impulses into developed stated policy. When you hear Rice speaking, that's what Bush would sound like if he were as articulate as Rice is."[71]

Their current voyage, of course, began where it did for most of us—on September 11. The key to Bush's first term was his decision, in the days after the attack, to channel his policy, and our national anger and resolve, to the task of combating terrorism and the nations that sponsor it all over the world—rejecting the narrower mission of just rounding up and punishing the particular al Qaeda operatives who planned 9/11. According to *Newsweek,* it was Rice who helped the president respond this way to the terror attacks: "Bush's moral impulses were easier to channel after 9/11. Rice was one of Bush's advisers who instantly saw that the War on Terror was global."[72] As Colin Powell, then Bush's secretary of state, told *Newsweek,* "The initial knee-jerk reaction after 9/11 was to go after al Qaeda." But Rice encouraged the president to focus on state sponsorship of terrorism as well. When Bush used the phrase "Axis of Evil" in his State of the Union address, it was an echo of what Rice had been telling him since the week of 9/11.

In a sense, Bush and Rice had both come a long way from their starting points. President Bush evolved from his father's perspective on foreign policy, with its emphasis on preserving the balance of power. Rice, too, had started as an apostle of the balance of power philosophy of Hans Morgenthau. But together they shaped a new consensus—a very Wilsonian worldview, based on universal values and a commitment to freedom and democracy.

It could have turned out differently. The Bush camp circa 1999–2000 was populated by the voices of "neorealism." In the *New Republic,* Jacob Heilbrunn quotes a senior advisor to the Bush campaign as saying that "as power diffuses around the world, America's position relative to others will inevitably erode. . . . The proper goal for American foreign policy . . . is to encourage a multipolarity characterized by cooperation and concert rather than competition and conflict."[73] Heilbrunn also quotes the Bush aide as saying that "order is more fundamental than justice." But Rice prevailed over these dour and pessimistic voices and helped President Bush to a new, values-oriented optimism about America's global role.

In an interview with the *National Review,* Rice noted how different this worldview was from the prevailing wisdom in the more secular and anticlerical environment of Europe. "Power matters," Rice said. "But there can be no absence of moral content in American foreign policy, and, furthermore, the American people wouldn't accept such an absence. Europeans giggle at this and say we're naïve and so on, but we're not Europeans, we're Americans—and we have different principles."[74]

The Bush/Rice approach to international relations went beyond just seeing the War on Terror as global. Beyond a confrontation with the Axis of Evil countries that promoted terrorism, they came to realize that until the world is populated by democracies, the resulting frustration and anger would make peace and stability impossible. This broader consensus, which animated Bush's second Inaugural Address, began to appear more and more frequently in the remarks of both the president and of his national security advisor.

In 2002, Rice began to speak of "a balance of power that favors freedom," an interesting merger of the language of geopolitical strategy and the objectives of a morally based foreign policy.[75] In a June 2003 speech to the International Institute for Strategic Studies in London, Rice laid out the case for a freedom focus more elegantly: "To win the War on Terror, we must also win a war of ideas by appealing to the decent hopes of people throughout the world . . . giving them cause to hope for a better life and brighter future . . . and reason to reject the false and destructive comforts of bitterness, grievance, and hate." Terror, she told the group, "thrives in the airless space where new ideas, new hopes and new aspirations are forbidden. Terror lives when freedom dies. True peace will come only when the world is safer, better and freer."

In her remarks, Rice decried the impulse of our allies to check and limit the American emphasis on freedom: "Why would anyone who shares the values of freedom seek to put a check on those values? Democratic institutions themselves are a check on the excesses of power. Why should we seek to divide our capacities for good when they can be so much more effective united? Only the enemies of freedom would cheer this division. Power in the service of freedom is to be welcomed, and powers that share a commitment to freedom can—and must—make common cause against freedom's enemies."

Rice echoed this emerging theme—that the War on Terror must be a war for freedom—in her commencement remarks at Stanford in June 2002: "Yes, people want to be free from want and to escape daily struggle for survival. But this is not what stirs the human soul or bridges the seemingly unbridgeable cultural divide. That bridge is the burning desire for liberty. Given a choice between tyranny and freedom, people will choose freedom. People want the best for their children and they want their creativity and their hard work to be rewarded. People want the freedom to speak their minds, to choose those who will lead them and the right to embrace their faith."[76]

In a sense, Rice and Bush had arrived at a point of view that had been pushed by Dick Cheney since his days as George H. W. Bush's defense secretary. The increasing influence of Cheney's moralist point of view may have been due to a silent consensus that he had been right during the first Bush presidency. In any event, it seems likely that Cheney's views affected both George W. Bush and—perhaps through him—Rice herself.

By 2002, Rice was articulating the difference this way: "Realists play down the importance of values while emphasizing the balance of power as the key to stability and peace. Idealists emphasize the primacy of values and the character of societies as crucial to a state's behavior toward other nations."[77]

The consummate Cold War realist had come a long way.

Rice's tenure as national security advisor, of course, was marked by extraordinary challenges—and no small share of criticism. Among the charges leveled against her, the one that merits consideration is that she failed to take action to avert the attacks of September 11.

As the Bush administration prepared to take office in early 2001, outgoing Clinton administration officials warned their successors of the specific danger Osama bin Laden posed. Despite their failure to kill or arrest him—and the many false starts and errant steps they took when he was vulnerable—Clinton's national security advisor, Sandy Berger, insists that he emphasized bin Laden's threat forcefully in conversations with Rice. And there were real-time warnings as the months ticked by: Rice herself reported, in her testimony before the 9/11 Commission, that President Bush "received . . . more than forty briefing items on al Qaeda" before September 11.[78]

Rice notes that President Bush "was tired of swatting flies" by responding to each terrorist attack and that he was moving to create an overall strategy to disable al Qaeda. The result was Bush's first national security policy directive—a plan to eliminate al Qaeda, published on September 4, 2001, only a week before the attack.

Should the administration have moved more quickly? What information was available to them? While reports of al Qaeda activity increased throughout the spring and summer of 2001, Rice testified, "the threat reporting that we received . . . was not specific as to time, nor place, nor manner of attack. Almost all of the reports focused on al Qaeda activities outside the United States, especially in the Middle East and North Africa."

Rice's critics have often cited the August 6, 2001 president's intelligence briefing as a smoking gun proving that she and Bush had advance warning of 9/11 and failed to respond. The briefing highlighted the possibility of terrorist attempts to hijack American aircraft—but, as Rice points out, it "did not raise the possibility that terrorists might use airplanes as missiles," and "referred to uncorroborated reporting from 1998 that terrorists might attempt to hijack a U.S. aircraft in an attempt to blackmail the government into releasing the U.S.-held terrorists who had participated in the 1993 World Trade Center bombing."[79]

Even if Bush and his administration had received certain warning of the 9/11 attacks, however, whatever efforts they could have taken to avert it would have been hampered by the mistaken decisions of the Clinton administration.

In our book *Because He Could,* we spell out in detail the wrongheaded judgments of the Clinton era that left us vulnerable on September 11:[80]

- In 1994, the FAA specifically changed its policy to allow travelers to carry small knives and box cutters—the weapons used on 9/11—on board American passenger aircraft.
- Because no one seems to have considered the possibility that terrorists might mount a suicide attack, the otherwise successful profiling system Al Gore designed to spot likely terrorists was of no use on 9/11. While it identified eleven of the nineteen hijackers, they were allowed to board the planes because it was thought that they would never blow up a plane on which they were traveling.
- The Department of Justice barred the FBI from examining the computer of Zacharias Moussaoui, the alleged twentieth hijacker, because of their concern that the information was obtained without a search warrant.
- Three previous attempts to kill bin Laden were scotched by the Clinton administration, one because of fears that the terror leader might be hurt or killed and the United States accused of using assassination as a tool of public policy.
- A fourth attempt to kill bin Laden by missile attack was cancelled, despite a very high chance of success, because the United States had just mistakenly bombed the Chinese embassy in Belgrade, and the administration worried about being accused of being bomb-happy. Also, Clinton was seared by allegations that he had tried to use a previous missile attack on bin Laden to divert attention from the impeachment inquiry.
- The administration had rejected proposals to ban the issuing of driver's licenses to illegal immigrants and to interface FBI and motor vehicle records to spot terrorists on watch lists when they were picked up at routine traffic stops. Because of this policy—inspired by concerns over privacy—three 9/11 hijackers, including leader Mohammed Atta, were released after being apprehended by traffic police.

In the face of such a massive record of incompetence throughout the 1990s, it is simply unrealistic to blame George W. Bush and Condoleezza Rice for failing to avert 9/11. The fact is that, with her guiding hand, the Bush administration was taking systemic, serious-minded steps to address the threat—steps made necessary by the halfhearted approach of the Clinton administration, which amounted to fiddling while Rome prepared to burn. The blame for our vulnerability must be laid at Clinton's feet, not those of Bush or Rice.

THE SPIRITUAL SIDE OF LEADERSHIP

Religious values are at the core of the Bush-Rice approach to global politics. If Bush's spirituality has been tested in the crucible of American politics, Rice's met challenges from the secular world of Stanford in which she lived before her return to Washington. As B. Denise Hawkins writes in *Today's Christian,* in her years as provost "Rice found that her colleagues' skepticism about religious belief was at times challenging. But that was not a hindrance. In fact, she says defiantly, 'I've been totally unflappable in my religious faith, and believe that it is the principal reason for all that I've been able to do. My faith in God is the most important thing. I never shied from telling people that I am a Christian, and I believe that's why I've been optimistic in my life.' "[81]

Like George W. Bush, Rice relies on prayer as she ponders foreign policy questions. As she explains it, "Prayer is very important to me and a belief that if you ask for it, you will be guided. Now, that doesn't mean that I think that God will tell me what to do on, you know, the Iran nuclear problem. That's not how I see it. But I do believe very strongly that if you are a prayerful and faithful person, that that is a help in guiding us, as imperfect beings, to have to deal with extremely difficult and consequential matters."[82]

Rice is careful to avoid a messianic approach to her mission, focusing instead on keeping a humbling sense of doing the Lord's purpose. She seems well aware that arrogance can leave one oblivious to reality, while humility offers a chance to see the world clearly and from a broad vantage point. As she puts it, "I try always to not think

I am Elijah, that I have somehow been particularly called. That's a dangerous thing. In a sense, we've all been, to whatever it is we are doing. But if you try to wear the imprimatur of God—I've seen that happen to leaders who begin too much to believe—there are a couple of very good anecdotes to that. I try to say in my prayers, 'Help me to walk in Your way, not my own.' To try to walk in a way that is actually trying to fulfill a plan and recognize you are a cog in a larger universe."[83]

Rice offered a window into her approach to adversity—and she had her cup full on 9/11—when she preached on "The Privilege of Struggle" at the Menlo Park Presbyterian Church soon after becoming provost at Stanford: "Struggle and sorrow," she said, "are not license to give way to self-doubt, to self-pity, and to defeat [but] an opportunity to find a renewed spirit and a renewed strength to carry on." Rice noted that it was through "struggle that we . . . get to know the full measure of the Lord's capacity for intervention in our lives. If there are no burdens, how can we know that He can be there to lift them?"[84]

For Rice, the War on Terror, with its apocalyptic start on September 11, brought a "certain moral clarity" to international affairs.[85] Rice explains: "I feel that faith allows me to have a kind of optimism about the future. You look around you and you see an awful lot of pain and suffering and things that are going wrong. It could be oppressive. . . . Then my only answer is it's God's plan. And that makes me very optimistic that this is all working out in a proper way if we all stay close to God and pray and follow in His footsteps."[86]

And if there were any doubt about the link in Rice's mind between her faith and her secular mission to advance democracy in the world, a glance at her remarks at a Sunday school class at the National Presbyterian Church in Washington would dispel them. Here, in this religious setting, Rice explained her commitment to freedom:

> I've watched over the last year and a half how people want to have human dignity worldwide. You hear of Asian values or Middle

> Eastern values and how that means people can't really take to democracy or they'll never have democracy because they have no history of it, and so forth. I remember all the stories before the liberation of Afghanistan that that nation wouldn't "get it," that they were all warlords and it would just be chaos. Then we got pictures of people dancing on the streets of Kabul just because they now could listen to music or send their girls to school.[87]

It may surprise some to realize that Hillary Clinton, too, is a spiritual person. And she's not afraid to integrate spirituality into her political vision. But a careful reading of her public statements suggests that sometimes her embrace of religious themes seems to serve as part of an elaborate defense mechanism to ignore the criticisms of her opponents.

In *Living History*, for example, Hillary recounts an encounter with South African leader Nelson Mandela in which he publicly thanked three of his former jailers for their kind treatment during his imprisonment. Hillary praises his "generosity of spirit," but then likens the "hostility" she faced in Washington over Whitewater, Vince Foster, and the Travel Office to the South African's persecution. She notes that "gratitude and forgiveness, which often result from pain and suffering, require tremendous discipline," and concludes that if Mandela can forgive his jailers, she too can try to forgive.[88] Oh boy! To equate charges of hiding billing records and making false statements to a special prosecutor to standing up for racial justice and being imprisoned for decades betokens so grandiose a self-image and such an elaborate matrix of self-justification as to be staggering.

Spirituality lies at the core of Hillary's liberalism. Her Methodist background is steeped in a tradition of public service and help for the poor. Noting that "I've always been a praying person," the senator has said that it is important for religious people to "live out their faith in the square."[89] At a time when the Democratic Party is relentlessly attacking the role of evangelicals in Republican politics, she boldly stakes out a position as a spiritual person, separating herself from the thrust of Democratic rhetoric. And this is no matter of political expedience: In private, Hillary is indeed a religious person with a serious faith.

But Hillary's politics sometimes seem at odds with the very values that she espouses. As first lady, Hillary admitted to "wondering if it was possible to be a Republican and a Christian at the same time."[90]

In Hillary's view, the ends always justify the means. Her commitment to helping the poor is, in her mind, so fundamental, so Christian, that it is permissible to deceive and mislead others to get the power to deliver on her priorities. So deeply does she believe that she alone has the answers that she seeks to impose her will on whatever political situation she faces. It is this deep belief in the ends justifying the means, for example, that led Hillary to hire detectives to track down and harass women who her husband had seduced, to achieve the higher goal of protecting his presidency. Or to persuade her husband to pardon the FALN terrorists in an effort to help her win the Hispanic vote in New York, a necessary step in the higher goal of electing Hillary senator.

Both women are animated by their religious backgrounds and faith. What direction that might take them in as president remains to be seen.

TWO WOMEN, TWO PATHS

A comparison between these two women, at parallel points in their lives, offers an instructive look at their respective choices and characters.

At the age of nineteen, for instance, Hillary Rodham was in her freshman year at Wellesley, unsure of her ability to compete in her new environment. "I didn't hit my stride as a Wellesley student right away," she writes. "I was enrolled in courses that proved very challenging. . . . A month after school started, I . . . told my parents I didn't think I was smart enough to be there." But she righted herself, and "after a shaky start, the doubts faded."[91]

At nineteen, Condi Rice was graduating Phi Beta Kappa from the University of Denver, the most honored female member of the student body. She already had achieved tremendous stature as a musician and figure skater and was looking forward to a career in international relations.

When Hillary was thirty-two, Bill Clinton began his service as Arkansas governor, and Hillary was promoted to partner in the politically connected Rose Law Firm, the most prestigious in the state. The connection between Bill's political rise and Hillary's legal career is undeniable. With no husband to latch on to, Rice earned a Ph.D. and a postgraduate fellowship and was promoted to associate professor of political science at Stanford. At thirty-five, as Hillary was settling into life as the first lady of Arkansas, Condi began her service on the National Security Council as the president's major expert on the Soviet Union.

At forty-six, Hillary became America's first lady, having survived the taunts and brickbats of her husband's race for president. Her first task? A disastrous effort to reform health care, which sullied Clinton's first two years in office and led directly to her party's humiliating loss of both houses of Congress—a loss from which it has not yet recovered.

At forty-seven, Rice was appointed national security advisor to President Bush; within months she was helping to shepherd the nation through the crisis that began with the attacks of 9/11.

Indeed, it was not until 2000 that Hillary Clinton stepped out on her own for the first time, running for the Senate. Then again, "on her own" is a bit of an exaggeration. With the president of the United States and all of his resources in her corner—raising millions for her campaign, providing invaluable research and policy guidance, assuring her the nomination without a primary in a state where she had never lived, and guiding her every move—

even on this first solo journey Bill was the wind beneath her wings.

It is difficult to tell, in assessing Hillary's career, where Bill ends and she begins; we are left with a list of joint achievements, and no real understanding of how to allocate them.

For Condi, we face no such difficulty. Her record is her own—from her early errors to the accomplishments she has accumulated in her most visible and important role. If Hillary's rise was based on political calculus, Rice's was made on her merits.

7

The Two Hillarys: Dr. Jekyll and Mrs. Hyde

There is, about Hillary Clinton, a quality that recalls the Robert Louis Stevenson novel *The Strange Case of Dr. Jekyll and Mr. Hyde*. Depending on the circumstances of the moment or her immediate political goals, Hillary shows us completely different, and sometimes conflicting, personalities and styles. The Hillary you see in public—the good Dr. Jekyll—is by no means the same acerbic Mrs. Hyde seen by her intimates in private.

Hillary has no intrinsic personality disorder. She is not schizophrenic; nor does she suffer from multiple personality psychosis. No, there is really only one Hillary: Mrs. Hyde. The other, Dr. Jekyll, is a contrivance, a persona carefully designed to present an idealized version of what she thinks the voters want in a woman candidate.

Dr. Jekyll is a polished, congenial, and moderate professional; Mrs. Hyde is an extreme and outspoken political ideologue. At almost all times and with great discipline, Dr. Jekyll is deliberately displayed, while Mrs. Hyde is hidden in the attic.

Her Jekyll and her Hyde don't have much in common. Hyde, the private Hillary, is often bitter and sarcastic, a partisan who is always looking for—and finding—enemies, plots, and conspiracies. She's the one who loves to study the charts her minions assemble, tracking the people and entities that make up her "vast right-wing conspiracy." She is sure that those enemies are hiding behind every door, waiting to pounce and destroy her and her husband.

But the public Hillary, Dr. Jekyll, has become the very symbol of bipartisan cooperation. She's the one who is openly and happily working with the same people who led the congressional impeachment drive against her husband, the people she once counted among her mortal enemies. The public Hillary chirps charmingly in interviews about the importance of bridging the partisan divide, regardless of history. Always smiling—even giggling loudly if there is a television camera anywhere in sight—she is the epitome of good humor.

But, at night, behind closed doors—usually at Democratic Party events—Mrs. Hyde comes out: the dark Hillary, furiously attacking the GOP and her right-wing enemies. Her rage is obvious as she attacks the president, his policies, and his administration.

Over the past six years, Hillary has undergone a calculating, audacious, and amazingly successful public makeover of not only her physical appearance, but her personality and politics as well. The goal of this sensitive but necessary transformation was to present a carefully packaged, attractive, and acceptable new Hillary to the American voters: the Hillary Brand.

The first thing to tackle was her appearance. Throughout her husband's presidency, Hillary's lack of any consistent signature style and her constantly shifting looks reinforced a public perception that Hillary was secretive, even untrustworthy. By the time she began seriously considering a Senate run in 1999, that perception needed to be changed. She was already inviting skepticism by running for public office herself and doing so in a state where she had never even lived. But she had been active in politics and on public policy issues for decades, especially as an advocate for women and children. Her campaign advisers began capitalizing on her history, working to strengthen the impression that Hillary was an experienced candidate who understood the issues and the process; their repeated mantra was that she had spent "thirty years fighting for women and children." But whenever her public record was reviewed on film, the first thing that jumped out at viewers (and voters) was that she never looked the same. At times, in fact, it was actually hard to recognize her: Many of her photos seemed to be of other people, not Hillary Clinton.

Throughout the Clintons' White House years, Hillary drew ridicule for her constantly changing hairstyle. One day she sported a flip, next day a French twist, then a ponytail, followed by a pouf or perhaps a bob. Short hair, long hair, blonde hair, brown hair; natural or highlighted, straight or curly—Hillary's ever-changing styles made her look a little fickle, but the eccentricity did the Clintons no great harm—until she faced the challenge of selling *herself,* not her husband, to the voters. She didn't look professional, didn't look like a serious candidate. She needed to look like a senator—and a senator from New York, not Arkansas. She needed a consistent image. Her handlers understood the urgency of stabilizing her looks. So, they came up with an acceptable new hairdo that would be seen every day. Since 1999, Hillary has always worn an attractive, stable, short, blonde bob. No more bad hair days, ever.

But the makeover went far beyond just Hillary's hair. Her wardrobe simply wasn't suitable for the image of a senator from New York. The jarring colors and strange styles, so foreign to the sophisticated Big Apple fashion world, were quickly relegated to the back of her closet. Gone were the turquoise short-sleeved suits and the garish plaid jackets, the hats and thick black stockings. In their place came the new Hillary costume: black pantsuit with alternating blouses of blue, pink, and white. Occasionally, for a slight change, there was a blue sweater tied around her neck. Day after day, throughout the campaign and for the next several years, Hillary wore the uniform. At night and for big occasions like the Democratic convention, she was bold and wore a turquoise pantsuit with matching contact lenses. Thus was born the look of the Hillary Brand—a well-groomed, professionally dressed, New York candidate. She was ready for prime time at last.

But Hillary's looks aren't the only thing that was transformed. Over the years, she's changed a lot more—her name, for example. In an awkward battle between her feminist self-image and the political needs of the moment, she's switched from Hillary Rodham to Hillary Clinton to Hillary Rodham Clinton; now, finally, she has become just plain Hillary. After the Clintons were married in 1975 and during Bill's first term as governor, she never used the Clinton name. This offended many Arkansans, especially when she sent out Chelsea's birth announcements with the parents named as Governor Bill Clinton and Hillary Rodham. In 1980, after Bill was defeated, in part because of voter animosity toward Hillary, she reluctantly became Hillary Rodham Clinton. Then in 1991, when her husband first ran for president, she dropped the Rodham and started calling herself Hillary Clinton. Once he was elected, she immediately made a formal announcement to the press that she was now to be called Hillary Rodham Clinton. Eight years later, when she ran for the Senate and wanted to establish herself as an independent politician and to distance herself from the tarnished Clinton name, she became "Hillary." No last name. Just plain Hillary.

And it may be that we haven't seen the last of her name changes. If Hillary is elected president, don't be surprised to see a press release explaining that, as of noon on January 20, 2009, the woman you expected to see become the second President Clinton will expect to be addressed as "President Rodham."

But the greatest metamorphosis in Hillary has been the change in her politics and her public persona. Like a fake chicken emerging from a plastic egg, yet another Hillary Clinton has surfaced. This one is the new Hillary Lite: a moderate, smiling, relaxed, open, honest, witty, chatty, charming, friendly, and easygoing New Yorker. Gone is the harsh, partisan, ideological, cold, calculating, Midwestern/Southern, ethically challenged woman we came to know in the White House. Instead, the confrontational liberal has been replaced with an accommodating moderate, right before our eyes. It's like watching a magician who pulls a rabbit out from under his hat: We know he's pulled a trick on us, but we can't quite figure out how he did it so quickly and deftly.

In *Rewriting History,* our 2004 rebuttal to Hillary's *Living History,* we catalogued the birth of the "Hillary Brand" and its role in Hillary's strategy of remaking and reinventing herself. But since then, the Hillary Brand has progressed far beyond her appearance, style, and need to connect with the voters. Beginning in early 2005, she has also undergone a serious, substantive change in her demeanor, her legislative initiatives, policy positions, and even her Senate allies. It's as if she made a New Years' resolution to-do list that included seriously changing her political image.

So it was that, at the start of this year, she began a concerted and blatant campaign to transform her public image from liberal to moderate. And she has successfully accomplished this by publicly speaking about important value issues, such as abortion, and framing her remarks in a way that is sympathetic to her opponents and even critical of her longtime supporters—who, like her, are adamantly pro-choice.

Like Dr. Jekyll, Hillary has put on a new public face of moderation, centrism, and religious values. The bad, old liberal Hillary Rodham—Mrs. Hyde, a reflexive liberal leftist like Ted Kennedy—still rears its head on unscripted occasions, but the kindly, reasonable Dr. Jekyll increasingly overshadows her.

Not that the real Hillary doesn't always agree with Ted Kennedy—she does—but she pretends not to so she can seduce swing voters. In order to do so, she must position herself in such a way that she will attract new supporters. She can't afford to look as doctrinaire as she really is. So instead, as the political winds change, she tacks widely to accommodate each new storm. This explains why she is incessantly shifting, continually changing—her rhetoric, her positions, her allies. It's all part of a grand scheme to elect Dr. Jekyll president in 2008 so that Mrs. Hyde can safely emerge. Whatever it takes, she'll do it.

Hillary learned, by watching her own husband's career trajectory, that liberals don't win national elections. The failures of McGovern in 1972, Mondale in 1984, Dukakis in 1988, and Kerry in 2004 underscore that liberals—like Leo Durocher's nice guys—finish last or, at best, second.

To understand the importance of moderation in winning presidential races, consider the contrast of Jimmy Carter and Bill Clinton in their respective campaigns for reelection. In their first attempts, both men were elected as moderates. Carter ran as the apostle of Democratic centrism, having made his national name by opposing McGovern's candidacy four years earlier and critiquing the party's drift to the Left. Bill Clinton ran in 1992 as a "new Democrat," committed to the death penalty, mandatory sentences for repeat criminals, a balanced budget, a work requirement for welfare, and a middle-class tax cut.

In their first years in office, however, both Carter and Clinton tacked to the Left, running away from the moderation that got them elected. They did so for the same reason: Democratic majorities in both houses of Congress moved them to the Left. But Carter stayed there, and when reelection time came, Ronald Reagan beat him. Clinton moved to the center in time for the 1996 election and beat his Republican opponent Bob Dole.

I stressed the importance of moving to the center to Hillary in July 1996, as we discussed whether the president should sign the welfare reform bill. From the first, Bill Clinton had supported requiring welfare recipients to work in order to get benefits and had backed time limits on welfare. But the Republicans kept giving him welfare reform bills he felt he had to veto. While these GOP bills embodied the welfare-to-work principle, they also failed to provide the adequate day care, food stamps, job training, and job-creation programs the president knew were needed if the process was to work.

But under the guidance of the new Senate majority leader—Trent Lott of Mississippi, who took office when Bob Dole left in May 1996 to run for president—the Republicans signaled a willingness to pass a bill the president could sign. Hurt politically by their intransigence in shutting down the government over the previous Christmas in a bid to force Clinton to accept their draconian budget cuts, the GOP needed to show some moderation to regain support. So Lott pushed through a bill with day care, job training, food stamps, and protective services for children—all the goodies Clinton wanted. But Lott insisted on including what most assumed would be a poison pill: cuts in aid to legal immigrants.

I worked hard to get Hillary's support for the legislation. In a funny scene told more fully in *Rewriting History,* I pretended to be a house painter and told Hillary that every four years I had to come in and do my job. To do it, I stressed, we needed to put the furniture in the middle of the room—that is, the center; once I'd left, I reassured her, they could put it back however they liked. Her reply? "You silver-tongued devil, you."

In her book *Living History,* Hillary shows that she got the message. "If he vetoed welfare reform a third time," she writes, "Bill would be handing the Republicans a potential political windfall."[1] As always, her calculations were political.

But how, you may be wondering, do we know that Hillary's periodic moves to the center aren't for real?

When you look at her record, one thing is clear: When it is important, she is a liberal. Her voting record is as left-leaning as possible. The *National Journal* evaluated Hillary's voting record and ranked her as the eighth most liberal member of the Senate—four notches more liberal than even Ted Kennedy and nine places to the left of Tom Daschle.[2]

She votes the party line on all judicial confirmations, supports all party filibusters, and never criticizes the Democratic Party when it is vulnerable. Her centrism smacks of tokenism and opportunism, limited as it is to legislation that is either optional or relatively uncontroversial.

It's astonishing that people have not seen through Hillary's transparent move to the center more readily; most observers appear to take it at face value. It sometimes seems as if she is testing the limits of Abraham Lincoln's famous dictum, "You can fool all of the people some of the time. You can fool some of the people all of the time. But you cannot fool all the people all the time."[3]

Again, Hillary's keen political calculus is responsible. Having begun her move to the center four years before the 2008 election, Hillary left herself plenty of time to persuade voters that she is the same creature of moderation that her husband was during the middle of his presidency. (Bill was very liberal before 1994, when he had to work with a Democratic Congress, and after 1997, when he needed the Senate Democrats to vote to kill impeachment. In between, he took a more moderate tack—and racked up most of the accomplishments of his presidency.)

According to a "Hillary meter," designed by pollster Scott Rasmussen, only 43 percent of Americans "view the former first lady as politically liberal." This figure has fallen from 51 percent at the end of January 2005.[4]

Despite her liberal record, though, as the new Hillary approaches her presidential candidacy, she is working hard to reposition herself in three vital areas. She is moving to the center of the ideological spectrum, aligning herself with the most ardent Republican right-wingers on uncontroversial and relatively unimportant issues; she is embracing religious and spiritual values more openly; and she is raising her profile as a hawk on terrorism and defense issues and as a knowledgeable expert on foreign affairs.

In these efforts, Hillary is really just borrowing from the triangulation playbook I helped her husband design as he faced the electorate in 1996. By embracing Republican goals of balancing the budget, cutting crime, and reforming welfare, he stole their thunder, drained them of passion, and set them up for the kill.

Now that she is running for president, Hillary is issuing a string of moderate policy statements on hot push-button issues, defusing Republican animosity by mirroring their positions. The one difference is that Bill Clinton was a moderate at heart, a politician who became a liberal only when he had to be.

On the dominant issue of our time—terrorism—Hillary hugs the hawkish line. She voted for the Iraq War, and though she criticizes the Bush administration for the way it is fighting the conflict, she constantly backs the war and votes for all the supplies, money, and troops Bush requests. In fact, she has called for the recruitment of 80,000 new soldiers in our army.

In staking out new ground for herself on national defense issues, Hillary has found a big ally: former House speaker and GOP stalwart Newt Gingrich. Hillary actively uses Newt as a stage prop to demonstrate her newfound political centrism. Serving together on an advisory panel on defense priorities, Gingrich and Hillary have gone out of their way to indicate a shared commitment to a strong defense. According to the *New York Times,* "Gingrich says he has been struck by how pro-defense [Hillary] Clinton has turned out to be at a time when other Democrats have criticized President George W. Bush's decision to go to war against Iraq. He chalked that up to her experience in the White House, where her husband had to deal with national security. 'Unlike most members of the legislature, she has been in the White House,' Gingrich said. 'She's been consistently solid on the need to do the right thing on national defense.'"[5]

For her part, Hillary relishes the relationship as a way to distinguish herself from her former liberal label, as she does all she can to heed her husband's advice and move to the center in time for the 2008 election. "I know it's a bit of an odd-fellow, or odd-woman, mix," Hillary said. "But the speaker and I have been talking about health care and national security now for several years, and I find that he and I have a lot in common in the way we see the problem."[6]

To understand just how contrived Hillary's sudden friendship with Gingrich is, one need only consult her own memoirs, published a little more than two years ago. Apparently, when writing the book, Mrs. Clinton forgot about her newfound admiration for the former speaker. Hillary mentions Gingrich twenty-five times in *Living History,* and not one instance is flattering. She makes no mention of the common ground she claims they now share. Instead she depicts Gingrich as a right-wing ideologue and enemy, the leader of the Whitewater attacks in Congress—along with her other new friend, former senator Al D'Amato. She criticizes Gingrich for erroneously claiming on *Meet the Press* that he didn't have government health insurance.[7] She blames him for protesters who appeared on her health care tour of the country.[8] She disparages him for his "glee" when the Republicans gained control of the House of Representatives in 1994.[9] She lambastes him for his proposal to set up orphanages for children of unwed mothers.[10]

Hillary also recounts how Gingrich's mother once told a reporter that her son had referred to Hillary as a "bitch."[11] After that, Hillary invited Gingrich, his wife, and his mother for a tour of the White House. In her account of the visit, Hillary returns Gingrich's insult by including a remark his then-wife, Marianne, made as Newt "pontificate[d] about American history."[12] "You know he will go on and on whether he knows what he is talking about or not," Marianne told Hillary.[13] And how did Hillary feel about being seated next to Gingrich at the congressional luncheon after her husband's second inauguration? "Perhaps it was someone's idea of a joke to seat me next to Newt Gingrich."[14] Hillary then outlines Gingrich's ethical problems; later, she makes sure to point out his own marital infidelity.[15] This is how she remembers her new friend.

Hillary's new alliances with the Right have not stopped her from attacking the president. When she criticizes President Bush, however, she tries to do so from the right, challenging his enormous federal deficit and calling attention to her husband's success in balancing the budget.

Hillary has also been turning heads when she rants about the dangers of illegal immigration. Support for immigrants—legal and illegal—has always been a chapter chiseled in stone in Democratic Party dogma, but now she is sculpting a new view. As the *Washington Times* has reported, "Hillary Rodham Clinton is staking out a position on illegal immigration that is more conservative than President Bush, a strategy that supporters and detractors alike see as a way for the New York Democrat to shake the 'liberal' label and appeal to traditionally Republican states. Mrs. Clinton—who is tagged as a liberal because of her plan for nationalized health care and various remarks during her husband's presidency—is taking an increasingly vocal and hard-line stance on an issue that ranks among the highest concerns for voters, particularly Republicans."[16]

In an interview on the Fox News Channel, Mrs. Clinton stressed her new hard line on illegal immigration, saying that she did not "think that we have protected our borders or our ports or provided our first responders with the resources they need, so we can do more and we can do better."[17]

On WABC radio in New York, she continued her new line: "I am, you know, adamantly against illegal immigrants." She added, "clearly we have to make some tough decisions as a country, and one of them ought to be coming up with a much better entry-and-exit system so that if we're going to let people in for the work that otherwise would not be done, let's have a system that keeps track of them."[18]

This is a new tune for Hillary to be singing: When the Clintons were in the White House, they were never interested in such programs because they were fearful of offending the Democratic Party base. In 1996, I urged the president to crack down on illegal immigration by denying driver's licenses to those who were not here legally. If our traffic cops could spot those who had overstayed their visas, I felt, we would have a potent tool against both terrorism and illegal immigration. But in those days Mrs. Hyde was still running things; Hillary opposed the proposal and helped to persuade her husband not to submit the legislation.

After 9/11, of course, the proposal was revived with a new sense of urgency, since the hijackers had more than sixty different driver's licenses among them, most obtained on false pretenses.

The proof that Hillary is being disingenuous in her newly vocal opposition to illegal immigration came in May 2005, when the proposal to ban driver's licenses for illegal immigrants was passed. Hillary voted yes, but only because it was attached to a defense appropriations bill to pay for our troops in Iraq and Afghanistan. She said she was "outraged" that it was attached to the appropriations bill and not treated as a stand-alone piece of legislation. She called the proposal "seriously flawed" and questioned whether it would help to fight terrorism. "We are being asked to vote on the so-called Real ID legislation," she said. "Its supporters say it is supposed to make our country safer, but how do we know that?"[19] But Hillary was well aware that, had the legislation been submitted on a stand-alone basis, a Democratic filibuster would certainly have killed it. Only as part of an omnibus "must pass" bill could it have any chance of being enacted into law.

But it is on values issues that Hillary is doing the greatest flip-flop. Ever since exit polls in the 2004 election showed that many voters identified "moral issues" as their chief concern, Hillary has done her best to show greater sensitivity to religious issues in particular.[20] The *Times* of London summed up the situation nicely: "If there was doubt before, there can be none now. Hillary Clinton is positioning herself for a run for the White House in 2008. With her recent attention to God, the military, and opponents of abortion . . . pundits . . . have drawn the obvious conclusion that the New York senator aims to be the first woman to occupy the Oval Office."[21]

Hillary has spent a lifetime opposing the pro-life movement in the harshest possible terms. In her 2000 Senate race, Hillary declared: "I am and always have been pro-choice, and that is not a right anyone should take for granted. There are forces . . . that would try to turn back the clock. We must remain vigilant."[22]

Now, however, Hillary's vigilance appears to be succumbing to the greater forces of her ambition. She recently declared her "respect [for] . . . those who believe with all their hearts and conscience that there are no circumstances under which any abortion should ever be available."[23] She noted, for the first time in her career, that "there is an opportunity for people of good faith to find common ground in this debate." She now goes out of her way to support proposals that would require parents be notified if their minor child has an abortion. Of course, she never mentioned her support for late-term abortion—or her other party-line pro-choice votes.

She has also jumped off the Democratic Party reservation on the issue of the president's "faith-based" initiatives, saying that there was "no contradiction" between Bush's proposals and the first amendment guarantee of separation of church and state.[24]

Hillary's move to the center has occasionally taken her into areas that some would see as incongruous. Recently, the woman who defended her husband as he squirmed his way out of accusations of gross misconduct said that there is too much sex on television in America today. Hillary called the proliferation of adult content "a silent epidemic" and noted that "just a decade ago, we made great strides to keep children away from inappropriate material"—calling to mind her husband's advocacy of the V-chip to block programming inappropriate for children.[25] Citing new forms of communication, Hillary said that advocates of child-appropriate content needed to "keep up with this multi-dimensional environment." She added, "All of us need to rise to this challenge."

Ironically, Hillary's Dr. Jekyll performance wins rave reviews precisely because of its contrast with the mean old Mrs. Hyde she used to be.

Hillary arrived in Washington to begin her Senate career with the most awful image that any new senator has had since Huey Long of Louisiana graced the chamber. Senators of both parties—like Americans generally—saw her as starchy, doctrinaire, highly partisan, and rigid, and many had doubts about her integrity. It was a reputation that was well earned during her White House years, and she knew she had to live it down when she entered the U.S. Senate. Only days into her term, she was forced to hold a press conference to defend huge payments made to both of her brothers from persons seeking—and receiving—presidential pardons from her husband. In those early days, her polls plunged: Her favorable rating among American voters declined from 52 percent when she took office in January 2001 to only 39 percent seven weeks later.[26]

As the *New York Times* put it, "For eight years, Mrs. Clinton had a reputation within some Washington circles as aloof and imperious, a reputation that—no matter how unfair her friends and advisers said it was—followed her right into the Senate after her election."[27] But then Washington found out that it was not Mrs. Hyde who was unpacking her bags as she moved into the Senate but a new creature: Dr. Jekyll. The *Times* noted: "Over the last year, Mrs. Clinton has surprised many of her colleagues with a series of personal gestures that have served not only to soften her image but also to help her fit into the clubby world of the Senate, where schmoozing and one-on-one politics go a long way."

Hillary cleverly did everything she could to moderate her own stereotype. Anyone who expected a pushy, attention-seeking prima donna in Senator Clinton was disappointed. For the first several years, with very few exceptions, she patiently and quietly learned the ropes and resisted the impulse to capitalize on her fame. She was deferential to other senators, did not hog the press, and made a lot of friends because of this posturing.

She disarmed her Senate colleagues with her openness and refusal to put on airs—pouring tea and coffee for them at committee meetings, inviting many of them to dinner at her Washington home.[28] She tried to show a softer, more feminine side. When Republican senator Kay Bailey Hutchison of Texas adopted an infant girl, Hillary gave her a baby shower. She called her own donors and asked them to help out her fellow Democratic senators. When New Jersey Democratic senator Robert G. Torricelli was facing a federal criminal investigation—

something with which Hillary has had broad personal experience—she consoled him.

She even charmed Senator John Ensign, an impressionable young Republican from Nevada, the land of second chances. The *New York Times* quoted him as saying, "She didn't come in with this air of superiority. She has a very warm way about her. You should see the looks I get when I tell people I like her," he continued, referring to voters in Nevada. "It's a look like, 'You've lost it.'"[29]

Another western Republican, Senator Craig Thomas of Wyoming, said that "the Hillary Clinton he had heard about is different from the Hillary Clinton he has met."[30] He said with amusement that she gave him a block of cheese from New York for Christmas. "I would have imagined that she would be a little pushy," Mr. Thomas said. "Based on what some of us had perceived, she has done very well."

The *Times* quoted one of Hillary's advisors, who said, in a classic understatement, "I think she is benefiting from low expectations."[31]

Hillary has also succeeded in using her newfound personal warmth, in the words of a 2005 *New York Times* story, "to cultivate a bipartisan, above-the-fray image that has made her a surprisingly welcome figure in some New York Republican circles, even as she remains exceedingly popular with her liberal base."[32] The paper noted that its poll showed that 49 percent of New York State Republicans approved of the job she is doing as compared with 37 percent who approved of it two years before.

A big part of Hillary's success in wooing New York State Republicans is her ability to play off her previous image as a partisan ideological extremist. The *New York Times* reported that Republican congressman Peter King—one of the handful of GOP congressmen that voted not to impeach President Clinton—said that "Mrs. Clinton had been anything but the liberal extremist that her conservative critics accused her of being. I'm not going to vote for her and probably disagree with her on 70 percent of the issues," he said. "But I think that too many Republicans who criticize Hillary Clinton sound like Michael Moore criticizing George Bush."[33]

Hillary can be exceedingly charming when she wants to be or has to be. In 1994, we hosted a fund-raiser for a Democrat running for Congress in our part of Connecticut. Hillary agreed to appear as the guest speaker. When I said that, everyone was excited about welcoming her to our home; Bill had visited us at our New York home, but Hillary never had. "Should I bring a covered dish?" she chirped lightly.

But Hillary is not *that* friendly or *that* moderate or *that* convivial or *that* reasonable. She benefits *now* from how bad her image was *then,* during her White House years.

Then, she faced daily criminal investigations as scandal piled upon scandal. At times she seemed to spend each new day defending herself against some accusation, all centered on her and her conduct. The Whitewater affair, the secrecy of the health care task force, the Vince Foster suicide, the removal of files from Foster's office, the firing of the Travel Office staff, the missing billing records about her work for the Madison Bank, her futures-market winnings, the FBI files discovered in the White House, the pardons her husband granted to Puerto Rican terrorists as she ran in New York State, her brothers' paid and successful efforts to secure presidential pardons for their clients, her $8 million book advance contracted days before she took office in the Senate, and the gifts she pilfered from the White House: All of these darkened Hillary's days before her Senate career.

Now, by contrast, there are no ethical questions to haunt her. Out of the limelight, she is also out of prosecutors' line of fire. No longer do her scandals dominate the front pages of the nation's newspapers; no longer is the nightly news filled with images of her parading into a federal grand jury or feeding stock denials to the press.

Then, Hillary came across as acerbic, partisan, and shrill. She blamed opposition to her health care proposals on big insurance companies, citing the same vast right-wing conspiracy she would later make famous during Monicagate. Hillary developed a well-earned reputation for slashing and burning her adversaries in political combat.

Now, conscious of her terrible partisan image, Hillary works overtime to cultivate Republicans and to speak well of them publicly. She comes on to them politically, focuses her considerable self-discipline on conjuring up a semblance of charm in the service of this New Hillary.

THE UNWELCOME REAPPEARANCES OF MRS. HYDE

Hillary has reformed her image, but Mrs. Hyde still lurks beneath the surface. Take, for example, her appearance at a June 2005 Democratic fund-raising luncheon in New York. Gone was nice old Dr. Jekyll; now, the Hillary of old reemerged to criticize the president in a loud partisan attack. "There has never been an administration . . . more intent upon consolidating and abusing power to further their own agenda," she told a cheering audience of party faithfuls. She lamented that "it is very hard to stop people who have no shame about what they're doing. . . . It is very hard to stop people who have never been acquainted with the truth."[34]

And Mrs. Hyde had no use for the spirituality of which Dr. Jekyll loves to boast: "Some [Republicans] honestly believe they are motivated by the truth, they are motivated by a higher calling, they are motivated by, I guess, a direct line to the heavens."[35]

It wasn't the first time Senator Clinton had fallen back into her old ways. A year earlier, on June 30, 2004, Mrs. Hyde had made an appearance before a Democratic audience in San Francisco and set forth her ultra-left fiscal and social philosophy. "Many of you are well enough off that . . . the tax cuts may have helped you," she began. "We're saying that for America to get back on track, we're probably going to cut that short and not give it to you. We're going to take things away from you on behalf of the common good."[36]

It was a rare glimpse behind the facade: At her core, as this soundbite confirms, Hillary believes in income redistribution—a policy of taking from one class and giving to others. I had a glimpse of her priorities when we discussed reducing the capital gains tax during the Clinton administration. I pointed out to the first lady that the tax made no money. If it were reduced, most economists agreed, the cut would stimulate the sale of assets, which would actually generate more, not less, revenue. (When the tax was cut in 1997, that is precisely what happened.)

Hillary argued against the cut, saying that it made sense to tax the wealthy. It was important to tax money earned by investment just as heavily as that earned by labor, she maintained. I pointed out that investors had already paid taxes on the money they were investing. But I focused even more on the key argument: that we could appeal to conservative voters by adopting a Republican program and use the money the tax cut would generate to avoid deeper cuts in spending programs.

But she wasn't buying it. It was too important to take from the rich even if it left you with less to give to the poor—a philosophy that would have bankrupted Robin Hood.

Sometimes her Phantom-of-the-Opera mask drops to offer a view of the real, raw, litigator/partisan underneath.

On May 16, 2002, eight months after 9/11, Hillary took to the floor of the Senate brandishing the front page of that day's *New York Post*. The paper's cover story reported the existence of the August 2001 intelligence briefing warning of the threat of terrorist hijackings of passenger aircraft. The 9/11 Commission, investigating this memo, reported that, while it might have heightened the president's awareness of the terror threat, it hardly amounted to actionable intelligence that might have prevented 9/11.

But Hillary jumped to a different conclusion. "The president knew what?" she shrieked. "My constituents would like to know the answer to that and many other questions. . . ."[37]

She continued: "The pain of 9/11 . . . is revisited today with the questions about what might have been had the pieces of the puzzle been put together in a different way before that sad and tragic day in September. . . .

"As for the president, he may not be in a position at this time to respond to all of those concerns, but he is in a position to answer some of them, including the question of why we know today, May 16, about the warning he received. Why did we not know this on April 16 or March 16 or February 16 or January 16 or August 16 of last year?"

Like a lioness returned to her cage, Hillary must have realized that her outburst went over the top; she never again repeated either her angry words or her enraged tone on that subject.

But there were other outbursts. At a closed dinner in Connecticut, Hillary—who didn't know there were reporters nearby—appeared to have what the *American Spectator's* David Hogberg called "a near meltdown," which Sean Hannity delights in playing on the radio. "I am sick and tired," she screeched, "of people who say that if you debate and you disagree with this administration, somehow you're not patriotic, and we should stand up and say we are Americans, and we have a right to debate and disagree with any administration."[38]

These days, however, such outbursts are the exceptions, not the rule. Hillary is usually extremely disciplined, and her handlers keep her carefully under wraps. Everything possible is done to ensure that she makes no mistakes. Nothing, no matter how small, is left to chance. In order to protect her from herself, everything that she says is completely scripted; she assiduously avoids any extemporaneous or spontaneous speaking at all—not because she is inarticulate, but because she gets into trouble and creates a record when she deviates from the message of the day.

Hillary's handlers are well aware that Mrs. Hyde has an unfortunate history of exaggerating and fabricating stories that have later caused her political problems after they were exposed. Such incidents usually occur when she is trying to connect with her audience and pretend she's just like them—trying hard to be normal. Hard-pressed to come up with actual experiences, she makes up stories she thinks they'll understand. Or she embellishes a real story, enhancing its details in order to elicit sympathy and paint her as just a regular parent, or wife, or daughter.

Sometimes her inventions are harmless—as with her concern one night when Chelsea was late getting back to the White House. Most mothers of teenage daughters would be relaxed if they knew that their daughter was protected by an armed Secret Service entourage in a bulletproof car that was constantly monitored and available by cellphone—and of course Hillary was never truly worried. But she wanted to be perceived as just another mom concerned about her teenage daughter. This tall tale was silly, but harmless; when she made up a similar story about her daughter's whereabouts on 9/11, on the other hand, it was tasteless at best, reprehensible at worst.

If Dr. Jekyll controls the microphone at press conferences, Mrs. Hyde is still the one casting the votes in the Senate—as Hillary showed when she toed the party line on filibusters of judicial confirmations.

In May 2005, the Senate came close to civil war over the issue of President Bush's judicial nominations. From the beginning of the Bush administration, Democrats had repeatedly used the filibuster to kill the president's nominations for judges in the U.S. District Court and various Federal Courts of Appeal. The word filibuster, which comes from the word "pirate," is a procedural mechanism that can be used to prevent debate on a bill—including judicial nominations. During the debate over the Civil Rights Bill in 1964, senators slept on cots and stayed in the chamber for days, reading out loud from books and magazines—anything to keep up the chatter in an attempt to stop the president's program to guarantee civil rights to all citizens. At that time, an actual live and constant debate was required. Once it stopped, there could be a vote on closing the debate and moving on to vote on the bill itself.

Today, the filibuster no longer requires the drama of round-the-clock debates. If the leadership knows that the requisite sixty votes don't exist to close debate on a particular bill, no debate is scheduled.

In May, the filibuster issue reached a head when Republicans threatened to change the rules to stymie the Democrats and prevent them from torpedoing the president's nominees. In response, the liberals promised that they would tie the Senate in knots through procedural wrangles if the GOP tried it.

Into this mess stepped fourteen courageous moderates, seven from each party, to bring sanity to the Senate.[39] These fourteen defused the conflict by postponing a rules change and allowing votes on three Bush nominees.

But when Hillary had the real chance to identify herself as a moderate in action, as opposed to words, she turned the other way. She was not part of the group of fourteen conciliators; in fact, she spoke not a word about the filibuster issue.

As anyone familiar with Hillary's public career knew, she'd faced this question before—and come down squarely in favor of the Republican objection to using the filibuster to kill legislation on partisan grounds. In 1994, when she wanted Congress to enact her health care legislation, she feared a filibuster would kill its chances. So she did her best to prevent such a tactic by trying to tack her bill onto a budget measure that could not be talked to death under the Senate rules. Her tactic was stopped by West Virginia's Democratic senator Robert Byrd, who had passed a rule requiring that amendments to a budget bill had to be germane.

So the record confirms that Hillary has done her share of flip-flopping. But isn't there a larger point here—that most people are comfortable in their own skin by the age of fifty-seven? Not Hillary Clinton. She's still searching for who she is—or, rather, for the woman she's supposed to be, the woman people will like, the woman voters will support. What's the right look, what's the right message? Even after eight years in the White House and five years in the Senate, even after the constant media exposure, she's still trying to reinvent herself into the symbol of the perfect female politician.

8

How Hillary Came to Be Hillary

Presidential candidates usually come from the ranks of the unknown or the vaguely familiar. Some, like Jimmy Carter and Bill Clinton, break into our consciousness from nowhere when they run for president. Others, like Ronald Reagan or George Bush, move from the edges of our political consciousness onto center stage. Still others, like Richard Nixon, come back from the dead and resurrect their careers promising a new approach.

But Hillary has been at the center of our national vision for fourteen years. She has become politically mature in the spotlight, like a greenhouse plant, drawing nutrition for her political growth from our attention.

We met her as the independent wife of the Arkansas governor. Then, as she campaigned for her husband, we saw her as the loyal helpmate who stood by her man. And, in the White House, we came to know her as Bill's alter ego—his surrogate pushing a failed health care reform program. Later in Bill's first term, she resurfaced as an independent star—flitting about the globe, writing books, and giving speeches. At the end of his term, she was the aggrieved wife, injured but not bitter, determined to fight impeachment while disciplining her husband privately. Finally, as she ran for the Senate, we came to know her as a political phenomenon without parallel in our era.

I first realized that Hillary might eventually run for president as I was meeting with Bill in September 1989. Then governor of Arkansas, Bill Clinton had tentatively decided to pursue the presidency in 1992 when the first President Bush came up for reelection. But the question that perplexed him was much more immediate: Should he run for re-election as governor when his term expired in 1990?

As Clinton pondered his decision, it became clear that if he didn't run, Hillary might go for the job in his stead. After polling the state, however, I advised the Clintons that Hillary was not likely to win. It wasn't that she was not popular. She was. But voters felt that her candidacy would be a ruse to keep the seat warm for Bill if the presidential race didn't pan out.

Hillary was deeply hurt. All her intense work reforming Arkansas education had netted her nothing: To the voters of the state, she was still just Mrs. Bill Clinton. The poll remained prominent in Hillary's mind as she made the transition from state to national politics during the 1992 campaign and its aftermath.

Burned by the 1989 survey, Hillary decided to chart an independent course as her husband moved onto the national stage. She would not repeat the mistakes of the Arkansas years. Nobody would see her as just Mrs. Clinton anymore. They would have to come to grips with her as a separate, independent political force—one who might, and probably would, seek office on her own, in her own name and on her own terms. Her determination was palpable: America would just have to get used to Hillary, not just to Hillary Clinton.

But the result was a disaster. From her very first foray on her own, Hillary crashed and burned.

Dragged into defending her husband from the accusations of Gennifer Flowers—which Bill eventually admitted to be true—she told the national media that she was an independent woman, not just some mouse who stood by her man and "stayed home and baked cookies."[1] The firestorm that erupted singed her and drove her back into the relative obscurity of behind-the-scenes campaign management, fund-raising, and carefully scripted campaign appearances. Independence would have to wait. Now she needed to get Bill elected.

After Clinton won, Hillary started looking for a new way to step out on her own. Should she be his chief of staff? Serve in his cabinet? Serve as secretary of education? Attorney general? How could she maintain her own independent image, get out from under his shadow, become a political figure in her own right, and not repeat the mistakes that had relegated her to Mrs. Clinton status for so long?

After learning that it was illegal for Bill to name her to a cabinet post (because of an antinepotism law passed after JFK appointed his brother Bobby attorney general), Hillary mulled the notion of heading a task force on a specific issue. The result, of course, was the total disaster of her health care reform initiative. Hillary tried to do too much, too fast and moved to expand government's role when people wanted it to shrink. Falling under the influence of a group of utopian liberals headed by Bill's buddy Ira Magaziner, Hillary became committed to the almost Marxist notion that you couldn't change one part of health care without changing the entire system.

But America wasn't buying it; Hillary's proposals met a solid wall of rejection in the Congress. Her effort to fly solo was decisively and humiliatingly shot down.

After she and Bill lost control of Congress in the 1994 election, the first couple asked me to come to Washington and help them out. One of my first tasks was to assess how Hillary was hurting or helping Bill. The public, I discovered, saw them as a zero-sum couple. The stronger she was, the weaker he seemed. The more the media wrote that Hillary was calling the shots, the more Bill seemed pathetically weak.

From these poll findings, a new Hillary emerged. No longer did she dominate White House strategy sessions. In fact, she stopped showing up. Instead, she devoted all her time to writing and traveling. No longer did she control—or, seemingly, even much care about—the White House agenda. That was Bill's job. Hillary had her own priorities.

She started writing a column, emulating her role model Eleanor Roosevelt. She collaborated with a ghostwriter to produce a book about education, *It Takes a Village,* that made the bestseller list for weeks on end. She traveled to seventy-eight foreign countries at taxpayer expense, usually with a retinue that included speechwriters, press people, makeup artists, and, often, Chelsea.[2]

Instead of generating press coverage and media attention by suggesting major new legislation or contesting issues in American politics, Hillary schlepped the press corps with her on trips abroad. That way she could maximize her media presence without generating the controversy that undermined her own and her husband's image. Hillary's handlers recognized that if a news organization paid for their reporters to accompany the first lady, they were going to use the stories they were handed.

Meanwhile, the president had reached his own conclusions in the aftermath of the failure of health care reform, and soon he was recasting his domestic agenda to move away from major legislative proposals. With Congress in Republican hands, he showcased a long series of minor policy proposals, which aggregated into major achievements and generated considerable political support.

These three strategies—writing, traveling, and focusing on narrow-gauge proposals—came to define not only the Clintons' approach to Bill's presidency, but also Hillary's own legislative approach in the Senate. Through two trips to Iraq and visits to Afghanistan and other global hotspots and by authoring memoirs that sold widely and well, Hillary has managed to stoke her image while avoiding serious issues and divisive controversy.

USING—AND DEFYING—THE FEMALE STEREOTYPE

Hillary is not simply defying the stereotype of liberal Democrats by trying to tack to the center. She is doing all she can to break through a more significant stereotype—the idea that women aren't comfortable authorizing the use of military force or balancing a budget.

It all began with a survey we conducted early in Bill Clinton's presidency to measure how women could and should run for public office. We interviewed two thousand voters in a national sample and divided them into four groups based on their gender and their support or opposition to the Equal Rights Amendment (ERA).

We asked voters, with all other factors being equal, whether a man or a woman in public office would do a better job of improving education, fighting poverty, being honest, and acting with compassion. Not surprisingly, women won all four categories. When it came to strengthening national defense, fighting crime, and holding down taxes, however, men came out higher.

What was particularly surprising was the fact that men and women, equal rights supporters and opponents, all shared the exact same stereotype. "It's not like racism," we told the first lady in a memo. "Voters who are racist will never back a black candidate while those who are not racist don't really care about race." But a woman candidate faced other problems: "The overt sexism in our country is less prevalent than racism," we wrote. "More people will never vote for a black than would refuse to back a woman running for office. But all voters stereotype women candidates. And every woman has to face the stereotype in a way minority candidates do not have to."

I briefed Hillary on the survey in the months after Bill's election to the White House. As I watch her strategies emerge today, it seems that my advice is serving as the cornerstone of her current presidential candidacy. "A woman candidate needs to go with the flow on the aspects of the stereotype that are favorable to her: education, poverty, compassion, children, integrity," I told her. "But she also must defy the stereotype by going out of her way to act tough on crime, defense, and taxes." I cited Margaret Thatcher and Israel's Golda Meir as examples of women who succeeded by defying the stereotype.

Hillary has apparently remembered another key finding in that poll. Her comments in a speech to a New York University forum about the integrity women show in public office echoed the poll results.

"Research shows the presence of women raises the standards of ethical behavior and lowers corruption," she told the group.[3] Having dodged grand juries and investigations throughout the 1990s, though, Hillary shows extraordinary optimism if she seriously believes that voters would necessarily conclude that her presence would raise the standards of ethical behavior in government. As I commented on a recent episode of *Hannity and Colmes,* "maybe women are more honest, but Hillary sure drags down the average."

During her tenure as first lady, Hillary worked hard to tap into the positive aspects of the female stereotype that our polling revealed. Her decision to focus attention first on health care, then on children's issues and education—culminating in the bestselling *It Takes a Village*—mirrored the priorities we identified in our polling three years before. In fairness to Hillary, these issues had always been important to her—but their importance was only redoubled by their marquee value.

As she now positions herself for her presidential candidacy, though, it is particularly impressive how Hillary has sought to emulate Margaret Thatcher in defying the negative stereotype of women as soft on defense. Even as Bill Clinton was criticizing Bush and the war in Iraq, Hillary Clinton was voting for the resolution authorizing it in the Senate. It cannot have been lost on either Clinton that the seminal moment in Clinton's presidential candidacy was when he said he would have voted to authorize the use of force in the first Gulf War in 1991. (Clinton told me that it was Gore's vote in favor of the war that militated in his favor when he was weighing offering him the vice presidential nomination.)

Shortly after taking her Senate seat, Hillary opted to join the Senate Armed Services Committee—a traditional home for hawks. As Michael Herson, a prominent defense lobbyist, puts it, "the Foreign Affairs Committee is basically focused on the diplomatic corps. The Armed Services Committee is the place to be in the Senate if you want to focus on foreign policy. They get into everything."[4]

Her membership on the Armed Services Committee permits Hillary to travel to military hotspots and be publicly seen with the troops. In 2003, she made a Thanksgiving trip to Afghanistan and Iraq—a gesture upstaged only by President Bush's own last-minute decision to fly to Baghdad himself. Coincidence? Hardly.

As her presidential candidacy draws near, Hillary is increasing her travels abroad, reinforcing her credentials as a foreign policy specialist while still battling the negative stereotypes of women in politics. In 2004, for example, she traveled to India where she spoke out about the India-Pakistani negotiations.

But her trips don't always work out as well as she would hope. In India, she defended American outsourcing of jobs—which benefits India enormously—and predicted that it would continue and grow. "Outsourcing will continue," Mrs. Clinton said in New Delhi. "There is no way to legislate against reality. . . . We are not in favor of putting up fences."[5] Hillary acknowledged the pressures to curb outsourcing: "I have to be frank," she said. "People in my country are losing their jobs, and the U.S. policymakers need to address this issue."

Unfortunately, Hillary was also a sponsor of a sense of the Senate resolution that expressed concern over the extent of job outsourcing and condemned the very practice she was endorsing in India.[6]

Hillary is outspoken on terrorism issues and has volubly condemned the Syrian presence in Lebanon. But she also criticized Ibrahim al-Jaafari, the new Iraqi prime minister—the first since free elections were held—expressing "concern" over his Shiite background and possible ties to Iran.[7] Here again, she put her foot in her mouth: Most observers have gone out of their way to underscore al-Jaafari's nationalist animosity to his coreligionists in Iran against whom Iraq waged war in the 1980s. Others have said that the last thing an American political figure should do is to muck around in Iraqi politics. And al-Jaafari himself answered that he didn't think he needed Hillary's permission to run and emphasized that she did not know what she was talking about.

Since 2005 began, Hillary has been acting like a shadow secretary of state—speaking out on every major foreign policy question. Can she be keeping her eye on Condi Rice? Or is she just doing what the 1993 poll told her to do?

The right answer, of course, is all of the above.

SURVIVING SCANDAL

Some people shake their heads in incredulity when they ponder a Hillary Clinton run for president, wondering how she can ever overcome all the scandals that plagued her during her White House years. But the fact is that she has. Hillary has been out of the White House for more than five years, and since she entered the Senate, the only allegations of impropriety leveled at her concerned her husband's pardons of donors, payments to her brothers from people who sought and received pardons, and her acceptance of expensive gifts and furniture. And all of these occurred before she entered the Senate.

But those are all history. They may be raised in a campaign, but they certainly don't define her in the years since she established her own political career. If the Republicans are to win, they must understand this and find issues other than decade-old scandals to run against her.

Just as tobacco companies can parade the warning labels they once resisted fiercely as proof that they had alerted smokers to the havoc cigarettes cause, so Hillary can point to the extensive Starr investigations leveled against her and her husband and say that she was never convicted or even indicted for anything.

Hillary can face the voters and truthfully say, "Kenneth Starr is no fan of mine. He did everything he could for five years to get me, and he came up empty. He investigated, hired the top detectives and lawyers, subpoenaed all kinds of records, even dragged me before a grand jury—and I never took the Fifth. And guess what? I was not only never convicted, but never even indicted, for anything." Indeed, Hillary can follow in the footsteps of the apocryphal criminal who ran for office after being acquitted on a technicality, declaring: "I've been proven honest." (And her good fortune has apparently rubbed off on those around her: When the national finance chairman for her 2000 Senate campaign was indicted by the Department of Justice for deliberately understating the cost of an August 2000 fund-raising event in Hollywood, he was acquitted by a jury in May 2005 after a one-month trial.)

So anyone who thinks that Hillary can be derailed because of her narrow scrapes with the law and her past ethical violations doesn't understand how skilled she is at avoiding blame and concealing evidence. She is a political Houdini, defying near-fatal political scandals time and again.

And yet, despite the obvious necessity of finding someone who can beat Hillary Clinton, skeptics still express doubt about the most obvious choice to oppose her. Could a political novice like Condoleezza Rice emerge from her cabinet role to win the nomination and the election? Yes, they say, her record is impressive. But . . .

9

But . . .

But there are still questions. Raise the idea of Condoleezza Rice running for president, and the same uncertainties come up over and over again.

Here, once and for all, are the answers.

BUT CONDI HAS NEVER HELD ELECTIVE OFFICE!

How serious an obstacle is lack of political experience when it comes to a presidential election?

Not much. Take a close look at our greatest presidents: Many of them did very well as president, despite a paucity of experience in elective office. In fact, a study of presidential biographies reveals that many of the most highly regarded presidents had performed very little service in elective office in the years before they became president. And what experience they did have was often very brief and very long ago.

Our greatest president, Abraham Lincoln, had almost no experience in elective office. His total national political service amounted to a single two-year stint in the House of Representatives—a term he served more than a decade before he ran for president. His congressional career, which lasted from 1847 to 1849, was scarcely relevant to his election as president in 1860. Except for that brief foray, his only experience in elective office was his service in the Illinois State Legislature—*twenty years* before he became president. Lincoln's real experience was as a utility lawyer, a profession that sharpened his skill as a public speaker and writer and led to his success in communicating with the American people before and during his tenure as president.

Three of our greatest presidents—Theodore Roosevelt, Woodrow Wilson, and Franklin Delano Roosevelt—had only the briefest exposure to high elective office before becoming president. Theodore Roosevelt served as governor of New York for less than two years before he was nominated to be vice president on a ticket headed by President William McKinley; he became president only a few months later, when McKinley was shot. Theodore Roosevelt's formative experience had nothing to do with elective office: He became famous as a lieutenant colonel of a regiment in the Spanish-American War nicknamed the "Rough Riders" and led them in a charge up San Juan Hill to rout the Spanish forces. Earlier, he had achieved some celebrity as New York City police commissioner. His experience in national office, like Rice's, was a matter of *appointive,* not *elective,* office: Roosevelt served as assistant secretary of the Navy, but for only one year, from 1897 to 1898. His only other electoral experience was a two-year stint in the New York State Legislature, twenty years before he became president.

Nor did elective office play much of a role in preparing Woodrow Wilson to be president. His reputation came from his years as president of Princeton University, where he became nationally known as an advocate of education reform. His record of raising academic standards at Princeton has interesting parallels with Condi's at Stanford. On the strength of his Princeton record, he was elected governor of New Jersey only a year and a half before he was nominated for president.

Franklin Delano Roosevelt had served only three years as New York's governor when he was elected president. While he had run for vice president twelve years before and served three years in the state legislature two decades before, neither experience was key to his ability to run for president. His most salient experience, particularly after World War II broke out, was the time he spent as assistant secretary of the Navy during the First World War—the same job his uncle Theodore Roosevelt once held.

From the more distant past, another great president, Andrew Jackson, had served only two years as senator from Tennessee before running for president. Jackson's main experience was military: He won the famous Battle of New Orleans against the British in the War of 1812.

So many of our great presidents have had little experience in elected office before they became chief executives. For most of them, their pre-presidential elective careers were incredibly brief and insignificant.

And some presidents—Dwight Eisenhower, Herbert Hoover, Zachary Taylor, and Ulysses Grant among them—had *never* served in elective office before becoming president.

Condoleezza Rice has served as America's principal advisor on our nation's main adversary in the closing years of the Cold War. She was the president's chief advisor as he navigated the difficult years after 9/11, and now she serves as secretary of state, piloting American foreign policy through an incredibly complex and dangerous world.

How would service in Congress or the Senate enhance her experience? Why should John Edwards be able to run for president without serious questions about his experience, despite the fact that he had served only one term in the Senate, while Rice, whose resume is so much fuller and more varied, is confronted with questions about her experience?

The experience that seems more fundamental than one's tenure in elective office is service in an administrative position. After all, the presidency is first and foremost an administrative job: The chief executive is responsible for overseeing the vast, sprawling executive branch and its hundreds of thousands of employees.

Since 1963, every American president—except Gerald Ford, who was not elected—has had extensive administrative experience before taking office as chief executive. Four served as governors before they became president—Jimmy Carter (Georgia), Ronald Reagan (California), Bill Clinton (Arkansas), and George W. Bush (Texas). Three were vice-presidents (Lyndon Johnson, Richard Nixon, and George H. W. Bush). Bush senior also had administrative experience as director of the Central Intelligence Agency and ambassador to the United Nations before he became president.

The candidates who have run for president without administrative experience during this period, in contrast, have fared very poorly. Senators Barry Goldwater, George McGovern, Robert Dole, and John Kerry all lost their bids for president.

While Hillary has held elective office, she has never had any administrative experience. In this area, Rice enjoys a decided advantage. Her six years as provost at Stanford, overseeing a billion-dollar budget and grappling with all kinds of administrative issues, has given her a wealth of background in the dynamics of management. As national security advisor, she also superintended a sizeable staff; her elevation to secretary of state put her in charge of the vast foreign affairs establishment, including 260 embassies, consulates, and other posts throughout the world with 30,266 employees, and a $10.3 billion budget.[1]

Hillary, by contrast, has no management track record at all. As an attorney at the Rose Law Firm, she was not a managing partner and had no staff to supervise. As first lady of the United States, she had a very small personal and political staff and faced no administrative challenges more profound than designing menus for state dinners and guest lists for White House receptions. On the one occasion when she did have a major administrative portfolio—her chairmanship of the Health Care Reform Task Force—she delegated most of the administrative and management duty to Ira Magaziner, and the result proved to be a total disaster.

In the U.S. Senate, Hillary oversees a staff of only about one hundred people—hardly a responsibility comparable with Rice's as secretary of state.

But beyond the issues of elective or administrative experience, isn't the real question how well each woman would do as president? How effectively could they handle the daily pressures of the presidency and the massive, unremitting public exposure that comes with the job?

Look at their records.

As first lady, Hillary Clinton carefully controlled her media exposure, shunning press conferences and allowing only carefully selected journalists to do interviews with strict guidelines to prevent them from straying into areas where she might be at a disadvantage. During her eight years in the White House, she was under constant fire for her role in scandals past and present. Her media strategy consisted of avoiding interviews and hiding from the press. She had only one difficult press conference: the so-called pink press conference, where despite the ceremonial trappings of the first lady's social role that surround her, she was forced to answer questions about some of her scandals. As chairman of the Health Care Task Force, Hillary was so secretive that she triggered—and lost—a federal lawsuit trying to keep the body's deliberations confidential.

As a senator, Hillary has continued to limit her public exposure to carefully planned speeches and other public appearances. She tends not to hold press conferences and is very selective about what questions she will answer. During her book tour to promote *Living History,* for example, she told one prominent radio talk show host that she would only do his show if he agreed not to ask her any questions about any of the White House scandals.[2]

Rice, on the other hand, has been in the public spotlight constantly, from her days as George W. Bush's campaign spokesperson on foreign affairs to her current service as secretary of state. She interacts with the media on a nonstop basis, opening herself up to any and all questions. In the global spotlight, her every comment is carefully scrutinized for its potential to create unwanted controversy or blow up in her face. And it never does. She has served for five years on the most public of stages without a single misstep.

Hillary, despite her efforts at media control, has had a long, long series of public utterances she would rather not have made:

- In the 1992 campaign, for example, she won national ridicule for saying she was a career woman who wouldn't "stay at home and bake cookies and serve tea."
- In 1994, her inept defense of her health care proposal led to its rejection in a Democratic-controlled Senate Committee, which refused to send it to the floor for a vote.
- Her efforts to explain the missing Rose Law Firm billing records became fodder for endless late-night comedians' jokes.
- Her attempt to blame the Lewinsky scandal on a "vast right-wing conspiracy" underscored her reputation for both paranoia and rabid partisanship.
- She was widely ridiculed for her claim that she was named after Sir Edmund Hillary, the hero who climbed Mt. Everest—five years after she was born.
- In the White House, she was embarrassed when a book revealed that she'd enlisted mystic Jean Houston to help her by "channeling" her predecessor, First Lady Eleanor Roosevelt, and asking her for advice.
- As she began her Senate bid in 2000, she had to explain comments she had made two years before, advocating a Palestinian state.
- She also had to spend a lot of time explaining away her kissing the wife of Yasser Arafat during a visit to the Middle East.
- Her claim to have always been a Yankee fan when she moved to New York to run for the Senate got quite a laugh.

After she became senator, the malapropisms didn't stop:

- In a speech in St. Louis, while praising Indian independence leader Mahatma Gandhi, Hillary made a racist joke, saying that Gandhi "ran a gas station down in St. Louis."[3]

- On February 24, 2004, Hillary said that women in Iraq were better off under Saddam Hussein than they are under the U.S. occupation. Under Saddam, she said, women "went to school, they participated in professions, they participated in the government and business world and as long as they stayed out of [Saddam's] way, they had considerable freedom of movement."[4] Hillary contrasted the situation with what she said prevails today, when "women tell me they can't go out" and face all kinds of restrictions. (Curiously, she didn't mention the hundreds of thousands of women Saddam murdered in his prisons and in poison gas attacks.)
- In April 2004, Hillary Clinton did an interview with a reporter for international distribution in which she criticized President Bush and the Iraq War. Her comments were reprinted around the Arab world. For example, Iran's news agency, Mehr News, quoted her as saying that "the U.S. is trapped in the quagmire of Iraq" and referring to the policies of the Bush administration as "arrogant and insolent."[5] She said that Bush was not "willing to admit his mistakes in Iraq, the grave mistakes that have endangered the lives of both the Iraqi people and the U.S. servicemen alike." She accused Bush of "endangering the peace and stability of the region." When Mrs. Clinton's remarks were broadcast by MSNBC's Joe Scarborough, Hillary's Senate office immediately denied speaking to the Arab press or criticizing the president. Scarborough says that "our producers immediately double-checked our sources and we tracked down the reporter who did the interview with Senator Clinton, and by noon we found out that the interview did take place. And the senator's press secretary then started to back down . . . and admitted that [Hillary] did the interview, did know the reporter was writing for an international audience and did criticize the president."

Compare Hillary's record of verbal gaffes and her difficulty at handling herself in public with Rice's flawless public appearance—and ask yourself which of these women is ready for prime time.

BUT CONDI HAS NO EXPERIENCE IN DOMESTIC ISSUES!

A more reasonable criticism of a possible Rice candidacy is her lack of a record in domestic issues. We really don't know much about how she would handle our most immediate concerns, and we can't consult a voting record to tell us where she stands. It is only in international affairs that she is a known quantity.

But many presidents have taken office with records heavily weighted to either the domestic or the international side of the ledger. It was said of Bill Clinton, for example, that his only experience in foreign policy was dining at the International House of Pancakes. And it was quite true that he had no background in foreign affairs, except for trade negotiations he conducted with other countries while serving as governor of Arkansas. On the other hand, Dwight D. Eisenhower only had experience in foreign affairs, having run the Allied command in Europe in World War II and served as NATO commander after the war. He took office without any grounding in domestic issues, but presided over America in the 1950s, one of the most prosperous periods of our history. President Harry Truman, who created NATO, the Marshall Plan, the Truman Doctrine, the Berlin Airlift, the policy of containment of the Soviet Union and fought the Korean War, had no foreign policy experience at all when he took office. Lyndon Johnson, whose foreign endeavors were much less successful, was also totally inexperienced in international relations. Nor did Jimmy Carter, Ronald Reagan, or George W. Bush have any background in global affairs when they became president.

Presidents who lack a strong background in either domestic or foreign affairs educate themselves about their fields of relative ignorance and often perform better than their more experienced colleagues.

But it is not accurate to say that Condoleezza Rice lacks a background in many of our most important domestic concerns. On the single most important domestic issue of the past five years—homeland security against terrorism—few can rival Rice's expertise. Her vast experience with national security problems surely better qualifies her to keep us safe and protect us against another 9/11 than the record of a mere U.S. senator. By 2008, she will have spent eight years atop the national security food chain. She will have had access to a stream of the most highly classified intelligence and served as a key player in decisions concerning our response to terror threats at home. What could be a more pivotal experience in handling domestic issues than that?

And when it comes to Hillary's seminal issue of education, Rice's years as a professor and provost at Stanford give her a decided advantage over the senator from New York. Rice is an educator; Hillary just writes about education. Mrs. Clinton's only hands-on experience with education, besides teaching Chelsea, has been her efforts, not notably successful, to reform the Arkansas school system, one of the most backward in the nation. Rice, by contrast, was a college professor for a decade and a top administrator at a major university for six years more.

Hillary claims expertise in three principal fields: education, health care, and terrorism. Clearly, Rice beats her hands-down in the matters of education and terrorism. And shouldn't Hillary's "experience" with health care do more than anything else to disqualify her from being president?

In any event, in our shrinking world, the distinction between foreign and domestic issues is increasingly artificial. The concerns of the globe are the same issues we characterize as domestic concerns here at home. The two are as intertwined and inseparable as the double helix of DNA. Together, they determine what kind of world we will live in. For example, when tens of millions of jobs are dependent on exports to other nations, how can one distinguish international trade policy from the challenge of job creation at home? Aren't the management of our trade relationship with China and the correction of our massive imbalance between imports and exports central to reducing unemployment here in the United States? Is the outsourcing of American jobs a foreign or a domestic issue? It's obviously both.

Protecting American intellectual property abroad, for example, is critical to the development of a healthy economy at home.

Fiscal and monetary policy, which are crucial to the management of the economy, are global questions of the first order. Whether the dollar declines or appreciates against foreign currencies helps to determine the fate of American exports and our jobs. The flow of money across national boundaries overwhelms the weight of purely domestic transactions as surely as international affairs override insular domestic concerns.

Can the problems of poverty, unemployment, school overcrowding, and demands on our health system be separated from questions about illegal immigration and the maintenance of security along our borders? Would not a grasp of bilateral relations with Mexico inevitably teach a secretary of state much that she would need to know to understand the problems that grip Southern California, Arizona, and Texas?

And many of the issues Rice has dealt with in the Bush White House are simply global manifestations of what might be called domestic issues. Earlier in this decade, for example, she pressed the president hard for a major program to fight AIDS in Africa. Along with Secretary of State Colin Powell, Rice was crucial in convincing President Bush to set aside $200 million for the Global Fund to Fight AIDS, Tuberculosis, and Malaria.[6] The money is part of a $15 billion package designed to fight AIDS in the most devastated areas, especially southern Africa.

The job of secretary of state will bring Rice up against major environmental issues, as she deals with concerns about climate change and reconciles Bush's refusal to sign the Kyoto Accords on global warming with the increasing evidence of dramatic and ongoing changes in world temperature.

As she addresses the balance between human rights and the need to fight terrorism, we will get a sense of where she stands on the questions that have so occupied our nation in recent years. Similarly, as global population control issues clash with the aversion of the religious right to birth control and abortion, we will see much more of who Rice really is and where she stands on the crucial issues we face.

On a few key domestic issues, Rice has already spoken out. And her opinions seem to flow from her own background and her spiritual journey.

For example, she insists that she is "a Second Amendment absolutist" in opposing gun control legislation—a position that stems from her childhood experiences watching her father and the other men of the neighborhood patrolling the community to keep the Klan out. Rice has never forgotten the importance of having weapons for self-protection when the law looks the other way.

As a beneficiary of affirmative action, Rice has broken with President Bush to endorse race-based preferences in college admissions. In a written statement explaining her views in 2003, Rice said, "I agree with the president's position, which emphasizes the need for diversity and recognizes the continued legacy of racial prejudice and the need to fight it."[7] But then she said, "I believe that while race-neutral means are preferable, it is appropriate to use race as one factor among others in achieving a diverse student body"—a position at variance with Bush's opposition to race-based affirmative action. (As previously mentioned, however, Rice believes that affirmative action should only go so far as to open the doors of opportunity—not to ensure tenure for professors who have not proven themselves.)

On the politically charged issue of abortion, Rice identifies strongly with President Bush's "view that we have to respect and need to have a culture that respects life."[8] "We have to try and bring people to have respect for [the culture of life]," she says, "and make [abortion] as rare a circumstance as possible." Despite this view, she calls herself "mildly pro-choice" because of her concern "about a government role in this issue." She affirms, however, her "strong" support of parental choice and parental notification, as well as her opposition to late-term abortion. On the issue of federal funding for abortion, she comes down against it "because I believe that those who hold a strong moral view on the other side should not be forced to fund it."

BUT CAN A WOMAN—A BLACK WOMAN, NO LESS—WIN THE PRESIDENCY?

We won't really know how a black woman will fare in a national race until one runs. There are only a handful of black female members of Congress, and almost all of them represent districts where a majority of the voters are African American. No black woman sits in the Senate, and only one member—Senator Barack Obama of Illinois—is African American.

As we've seen, though, national surveys suggest that a woman can be elected president. And the very idea that two might run against each other underscores how ready the nation is. In April 2005, pollster Scott Rasmussen reported that 72 percent of Americans "say they would be willing to vote for a woman president . . . [including] 75% of American women and 68% of men."[9] But still Americans ask the question: Are we ready for a woman president? The reason we keep asking is that while 72 percent of us say we would vote for a woman, only 49 percent think that "most of their family, friends, and coworkers would be willing to vote for a woman to serve as president."

Americans, in other words, need to get with the program! You may wonder whether your neighbors think the nation is ready for a female chief executive—but if you are, your neighbors probably are too.

Support for a female president cuts across party lines. Among Democrats, 84 percent say they would vote for a woman, while 61 percent of Republicans said they would. (The huge publicity a possible Hillary Clinton candidacy has attracted may well have depressed the poll's findings about Republican support for a female president: Plenty of GOP loyalists might have answered "yes" if not for the possible endorsement such a response would give *that* particular woman.)

And we needn't rely on surveys to measure our willingness to support a woman candidate. Fourteen women currently serve as U.S. senators, from states as diverse as New York, Maryland, Michigan, North Carolina, Louisiana, Arkansas, Alaska, and Texas.[10] In Maine, Washington State, and California, both senators are women. Eight women are governors, and fifty-nine serve in the House of Representatives.[11]

But what about an African American? Can a black candidate be elected president? Even ten years ago, during the flurry of interest in Colin Powell's candidacy, the numbers suggested as much. Our internal polls in the Clinton White House showed Powell defeating the president by 52–44. Though the general decided not to run, he clearly could have been elected. Being black may not be a real bar to a Rice candidacy. Indeed, it may be a big asset. Apart from her ability to pry African American votes away from the Democrats, the enthusiasm for Powell in 1996 may reveal a deeper truth about race relations in modern America: A large number of white people may be eager to vote for an African American to expunge our national legacy of racism.

Imagine what a Rice victory would mean to every African American boy and girl in the United States! It would mean that there is no ceiling, no racial barrier they cannot surmount. It would send a message to all businesses, colleges, universities, boards of directors, and government agencies that tokenism is gone, and real racial diversity is here to stay.

If the civil rights movement of the 1960s was animated by the haunting lyrics and melody of the song "We Shall Overcome," electing Condoleezza Rice to the White House would send a very different message: "We *have* overcome." And that, apart from Condi's obvious merits as a possible president, might just be worth voting for.

BUT . . . WHAT ABOUT HER SOCIAL LIFE?

Asked about her life as an unmarried woman, Rice says, "I am a very deeply religious person, and I have assumed that if I'm meant to get married that God is going to find somebody that I can live with."[12]

Her nearest brush with matrimony arose from a very unexpected source: football. In her younger years, Condi's social life seemed to revolve around football. Introduced to the game by her father, Rice became an avid, and by all accounts highly knowledgeable, fan. And her social life, as Rice's biographer Antonia Felix notes, "revolve[d] around football." At twenty-three she dated a member of the Denver Broncos, and their romance blossomed into a serious relationship. According to Condi's friend Deborah Carlson, her beau was a "very major player" on the team. Her biographer recounts how "Condi socialized with the NFL wives, sat in the good seats at games with them and became a well-loved member of that intimate inner circle of the NFL." The couple eventually got engaged; Rice "picked out her wedding gown and started to work with her mother on the arrangements." Carlson told Felix that "she was seriously going to marry him."[13]

But then the couple called it all off—the engagement and the relationship. "I really don't know what happened," Carlson says. "It wasn't anything like a major blowup; I think they just got to the point where they didn't get along."[14]

Felix writes that "she would date more football players as well as men in other lines of work."[15] But no ring.

Will America accept a single woman president? Consider this: 42 percent of all adult women in the United States are single—and one American in five is an adult single woman. Why wouldn't they vote for one of their own?

BUT WHAT IF SHE SAYS NO?

Condi *has* said no. She *is* saying no. And she will *continue* to say no—at least until she changes her mind.

And she deserves to be taken at her word. She has "no plans" to run for president. She has "never wanted" to run. It's hard for her "to imagine" herself "in that role." She has "no interest" in running. She has "no intention" of running. In sum, she says she won't run.

Got it?

Rice probably deeply believes that she will not run for president in 2008. But, if the American people want her to run, they can draft her into the role . . . and thrust greatness upon her.

If they build it, she will run.

All her refusals really mean—unless she finally echoes Sherman, refusing to run if nominated or serve if elected—is that the impetus for a Rice candidacy must come from the voters, not from her.

Every candidate pretends that it is his supporters who impel him to step forward and seek the office. Ever since George Washington, each has indulged in the fiction that the office is seeking him, when it is transparently obvious that it is the other way around. Even as the candidate huddles with pollsters, strategists, media creators, direct mail consultants, field organizers, advance staff, schedulers, convention planners, fund-raisers, key donors, and party honchos plotting and planning a run for president, it is still the office that is seeking the candidate.

But Condoleezza Rice really means it. She's not running. After all, she does have a rather demanding day job. Clinton, Kerry, McCain, and other possible candidates will remain in the Senate—a part-time job at best and one that allows plenty of time to run. Neither Kerry nor Edwards nor McCain nor Lieberman thought it necessary to leave the Senate to run in 2000 or 2004. Giuliani and Edwards work in the private sector, with flexible schedules.

Condoleezza Rice, on the other hand, is secretary of state—no part-time job. Not only does her current position deny her any time to indulge in preparing a run for president, it leaves no untapped restless psychic energy for the task. More important, she cannot serve two masters. Either she is working for Bush and the nation trying to achieve his and our foreign policy objectives, or she is in business for herself running for president. This dichotomy is not simply a matter of appearances. She must, deep down, have only one goal: to serve her nation and the world as secretary of state.

Her supporters, on the other hand, have the freedom to plan a candidacy around her. If they want Rice to run and they put their minds to it—spreading the word, raising the funds, lining up the delegates, winning the primaries, and nominating her on the convention floor—she will be the Republican candidate for president in 2008.

The one thing they can't waste time doing is begging Rice to run. Nor can they be disappointed when she says no. They must be diligent suitors. Her backers have the right to insist and to make their enthusiasm for her known. In this era of self-serving, ambitious, self-seeking politicians, would not voters welcome a person who is just doing her duty, heeding the will of the people?

Does she want to run? She may well not. She's probably never seen herself as a politician. The only thing she really wants to run, apparently, is the National Football League. Asked if she wanted to be national security advisor should Governor Bush be elected president, Felix reports, she skirted the question but added that "anybody who knows me, knows that it's absolutely true" that her dream job was to head the NFL. She said that if Bush were elected but "the NFL job comes up, the governor is on his own."[16]

The only time Condi and running for Congress were ever mentioned in the same breath was when an elected office became an appointed one. In 1990, Pete Wilson, California's Republican senator, was elected governor. After Wilson resigned from his old job, he had to name a successor. (When a senator retires in the middle of his term, the governor of the state—in this case Pete Wilson himself—appoints his successor.)

As Wilson cast about for Californians to select for the job, the Bush White House sent in a suggestion: Condoleezza Rice. Condi reportedly made Wilson's short list for the position, but it went instead to state senator John Seymour. While Condi said she was not interested in pursuing the appointment, there is no record that she asked that her name be withdrawn either.

Rice came closer to running for elective office when California residents circulated a petition calling for the recall of their Democratic governor, Gray Davis. Among the names circulated of Republicans who might oppose Davis was Condoleezza Rice. An April 2003 poll by Sacramento Republican political consultant Ray McNally found that Condi would have defeated Schwarzenegger by 66–17 and would have beaten both leading Democratic candidates.[17]

And this time Rice seemed a lot more receptive to the idea of running. The *San Jose Mercury News* reported that "Rice, 48, has been coy about her intentions, but those who know her well say the state's top job is potentially appealing to her. She has not publicly expressed interest and has taken no visible steps to lay the groundwork for a campaign. But key Republicans said she has sparked optimism by not ruling it out."

Her Stanford buddy Coit Blacker told the San Jose paper that "She has never said to me, 'This [running for governor] is something I would like to consider. This is something I would like to happen.' . . . But coming to her as an opportunity is very, very different than actively pursuing it herself. If the circumstances are right, my own guess is that this is a challenge she might be prepared to meet."[18]

For her part, Condi made the same kind of coy denials she is now making about running for president. "It's not on my radar screen," she told a California group when she was asked about running for governor.[19] "I've got my hands full," she added. But no Sherman pledges here either.

Can it be done? Can a candidate be drafted in this era of ambitious politicians, obsessive in their self-promotion? Yes, it can. And not only that: The record of the 2004 presidential campaign indicates that such a movement might have a greater chance of success now than at any time since bosses dominated the selection process. In 2004, politics turned upside down, and the people became the active players—the very conditions that would be required for a presidential draft.

10

2004: The Year Politics Turned Upside Down

In 2004, the extreme sentiments and voracious enthusiasm on both sides of the campaign swept the political landscape like a prairie fire, ignited by the spontaneous combustion of the grassroots. It was a new form of politics, which altered the process and marked the end of the era of media domination: At last the voters transcended their roles as mere spectators and became full participants in the process of selecting candidates.

And it is these very changes that make a draft of Condoleezza Rice possible.

Before 2004, voters stayed in the stands and watched the political game unfold on the playing field in front of them. On television, they followed the conventions, the debates, the ads, and the soundbites and formed their opinions accordingly. Beyond good-natured barroom or dinner party conversation, their role was essentially passive. The only things that were asked of them were to answer pollsters when they called and to vote on Election Day. Like little nineteenth-century children, voters were to be seen but not heard.

The average person wasn't even expected to finance a presidential campaign. While direct mail solicitations deluged us all, campaign managers did not expect much more than a 2 percent response. It was up to the big boys—the five-, six-, and seven-figure donors—to pay for the campaigns of their favorites or their puppets.

Ever since John F. Kennedy hypnotized us with his good looks and eloquence, television has been the central element in our electoral process. In the media age, politics was something that happened on television. The venues of our politics have been televised debates, media advertising, sound bites, news coverage, tarmac press conferences at airports throughout the nation, and Sunday morning interview shows. The effect has been anesthetizing, numbing voters and reducing them to the level of passive onlookers.

JFK convinced us that he was young and Nixon was old during their television debates. LBJ made us see Goldwater as a dangerous warmonger with his infamous Daisy ad showing an atomic bomb blast, a commercial produced by media guru Tony Schwartz. Nixon revived his career by unveiling a "new" version of himself in carefully controlled, televised town meetings. Jimmy Carter's infectious grin and Ronald Reagan's aw-shucks charm came through clearly on the tube. When Bill and Hillary had to recover from the accusations of Gennifer Flowers, they did so in an interview on *60 Minutes.* And when Bush unexpectedly outperformed Al Gore in the 2000 presidential debates, it was television that brought the surprise into our homes.

But television news itself has always been manipulated by the elite print media that set its agenda. Newspapers like the *New York Times,* the *Washington Post,* the *Wall Street Journal,* the *Los Angeles Times,* and, lately, *USA Today,* generated the stories from which the network news organizations took their cues. And, every Monday, *Time, Newsweek,* and *U.S. News and World Report* would chime in, helping to set the week's political agenda with their feature stories. The journalistic elites who published, edited, and wrote for these outlets dominated politics.

Life in the White House during the media age revolved around attempts to manage the dialogue with the media. It wasn't Congress, the bureaucracy, or the courts that dominated the White House decision-making process, but the media. It was speculation on how the press and television would react that consumed the majority of the waking hours of the White House staff and consultants. Once, while returning from a speech to a group of students, Bill Clinton told me how he explained the government's operation. "They don't realize that the media runs the government," he commented ruefully. "They think I do."

As the media cemented its domination of politics, fund-raisers and donors—who could find or give the money to buy advertising on television—assumed critical importance. And so did political consultants, who would help the candidates decide what to say on television.

When I worked in the Clinton White House, my role included writing the president's daily statements to the media. I would compose only a ten-second sound bite that would be at the core of the statement. Then I would turn it over to the speechwriting staff. "Make the rest of the statement so boring and so dull that they [the media] have no choice but to cover my sound bite," I instructed.

A small group of unelected people—perhaps fewer than two hundred—became the permanent arbiters of the system during the media age.

The media itself had taken power from the political bosses, who had run politics ever since parties were invented by Thomas Jefferson, James Madison, and Aaron Burr in 1800 and refined by Andrew Jackson and Martin Van Buren in 1828. Until television took away their power beginning in 1960, it was the political bosses who selected party nominees, meeting in the proverbial smoke-filled rooms to make their decisions like Mafia dons dividing up territory. A few states permitted presidential primaries, but most, including many of the biggest ones, had no role for voters in candidate selection. The decision as to whom a state's entire convention delegation would back was usually made by one man (never a woman) who had all the power.

It wasn't until 1960, when John F. Kennedy gained the nomination by winning selected primaries, that voter choice even entered into the equation. (JFK thought he had to prove that a Catholic could win, so he took the unusual step of predicating his candidacy on his performance in the primaries.) In 1964, right-winger Barry Goldwater upended the Republican establishment by defeating New York governor Nelson Rockefeller in the California primary, guaranteeing him the GOP nomination. Primaries to select candidates became universal after 1968, when the Democratic Party nominated Vice President Hubert H. Humphrey, who had not entered any primaries. Antiwar supporters of senators George McGovern and Eugene McCarthy (and the former backers of the assassinated Bobby Kennedy) demanded reform and forced the party establishment to require primaries or caucuses in every state. The power of the bosses was over. Television was in charge.

In the media era, one could follow the advertisements of each campaign, the speeches of the two candidates, the press clippings from a handful of top newspapers, and the nightly news coverage on network television and get all one needed to know about what was going on. When I worked for Clinton on the 1996 election, I monitored these news sources, just like a surgeon who checks vital signs throughout an operation.

Every night I would get tapes of the three network news shows dropped off at my hotel room. I asked my staff to tally how many seconds were devoted to each candidate, on what topics, and what percent of the coverage was positive and how much of it was critical. I would get a daily summary of the front page of twenty-five regional newspapers and read the *New York Times*, the *Washington Post*, *USA Today*, and the *Wall Street Journal*. The staff that traveled with the president would send me excerpts of what he and our Republican opponent Bob Dole had said on the stump that day. I reviewed tapes of our advertisements and those of the opposition. And that was all I needed to know to gain an overview of the entire campaign.

They were heady days. It was fun. But it's over.

In 2004, the process suddenly became much, much more pluralistic. It wasn't just the candidates or the parties, or even their special interest front groups like the trial lawyers, the labor unions, the medical associations, or the business groups, that controlled the dialogue. Now they were joined by millions of others who used every tool at their command to communicate with the voters. No one person, no matter how large his or her staff, could possibly keep informed about even a part of what was going on in the election of 2004.

In 2004, the grassroots replaced the media establishment in the driver's seat of our political system. In that fateful year, Americans got up out of their armchairs, turned off the TV, and went outside to participate in the election process.

In retrospect, it will not be Bush's reelection that will resound through the ages when people think of 2004. Just as many consider the TV debates of the 1960 campaign as of more lasting significance than Kennedy's victory over Nixon, the political science experts of tomorrow are likely to see 2004 as the year when the media lost its monopolistic hold on the election process. Suddenly, amazingly, television became only *one* of the players in the political game.

First, there were the books. Fifty political books made the *New York Times* Bestsellers List in 2004. Ten of the top thirty bestselling nonfiction books, according to *Publishers Weekly,* were about politics and the presidential election.

In past campaigns, the books appeared after the ballots were counted. Theodore White's *Making of the President* series and others that followed its lead would analyze what happened after it was over. The lead-time publishers needed to bring out a book was too long and popular interest too limited to make books a key element in the rhetorical exchanges before the polls opened.

But in 2004, books from both sides of the political spectrum dominated the stores, the Internet, and the talk shows.[1] From the left came *America* by the *Daily Show*'s Jon Stewart, which sold 1,519,000 copies; *Lies and the Lying Liars Who Tell Them* by Al Franken (1,030,450); *Dude, Where's My Country?* by Michael Moore, published in October 2003 (806,892); *The Family,* a negative biography of the Bushes by Kitty Kelley (715,000); *Against All Enemies* by Richard A. Clarke (540,000); *Bushworld* by Maureen Dowd (282,000); *The Price of Loyalty* by Ron Suskind (178,000); *Will They Ever Trust Us Again?* by Michael Moore (235,000); *Imperial Hubris* by Anonymous (not the Anonymous who wrote *Primary Colors* in the 1990s) (210,000); *What's the Matter with Kansas?* by Thomas Frank (187,000); *Worse Than Watergate* by John Dean (169,000); *American Dynasty* by Kevin Phillips (168,000); and *Chain of Command* by Seymour M. Hersh (111,000). These liberal books sold a combined six million copies. But it would be fair to add to the list Bill Clinton's *My Life,* whose two million copies weigh heavily on the scales—especially considering the size of each copy!

But conservative authors were just as prolific. *Who's Looking Out for You?* by Bill O'Reilly (932,750 copies sold); *Unfit for Command* by John O'Neill and Jerome R. Corsi (814,000); *American Soldier* by General Tommy Franks (660,000); *Deliver Us from Evil* by Sean Hannity (527,000); *How to Talk to a Liberal (If You Must)* by Ann Coulter (445,000); *Ten Minutes from Normal* by Karen Hughes (321,000); *A National Party No More* by Zell Miller (172,568); *The Faith of George W. Bush* by Stephen Mansfield (151,000); *Rewriting History* by us (106,000); *Michael Moore Is a Big Fat Stupid White Man* by David T. Hardy and Jason Clarke (105,000); and *Why Courage Matters* by John McCain (103,000): a total of 4.3 million sold.

Some books were not merely big sellers, but news events in themselves. Richard Clarke's *Against All Enemies: Inside America's War on Terror,* published in the spring of the election year, helped to shape the public debate. As Clarke's book soared to the top of the charts, its attack on the Bush administration's War on Terror came to dominate the hearings of the commission charged with investigating the 9/11 attacks. First Clarke himself laid out the charges before the commission and a national television audience numbering in the tens of millions; then Condoleezza Rice, speaking for the administration, replied to his accusations. Clarke's book did more to set the political agenda for the Left than anything either Kerry or the Democratic Party was able to do.

Watergate-convicted felon John Dean entered the fray with a book harshly critical of the president that attracted a wide readership: *Worse Than Watergate: The Secret Presidency of George W. Bush.* And pacing the attack on Bush's economic policies was *The Price of Loyalty: George W. Bush, the White House, and the Education of Paul O'Neill,* in which the former treasury secretary savaged his ex-boss. But the Right wasn't napping. *Unfit for Command: Swift Boat Veterans Speak Out Against John Kerry* by John E. O'Neill and Jerome R. Corsi plagued the Democratic campaign and forced Kerry on the defensive after his convention adjourned. Sean Hannity's *Deliver Us from Evil* made number one on the *Times* list with its strong attack on terrorists and their liberal apologists. And the conservative movement's sharpest tongue—and wit—brought her own unique style into the fray with *How to Talk to a Liberal (If You Must): The World According to Ann Coulter.*

Never before have the bookstore shelves so sagged under the weight of partisan literature in the months before an election. It was almost hard to wade through them to find the more usual literary fare of thrillers and diet books.

Why was everyone snapping up these volumes? To arm themselves for the next day's helping of website warfare.

In the chaotic, vibrant world of the Internet, partisan arguments raged throughout the runup to the election—millions of them, every day. Chat rooms became debating centers. Internet bulletin boards were crammed with messages on the election. Each day I would receive at least a dozen e-mails from each side featuring their arguments and showcasing the stories favorable to them.

NewsMax, the online wire service for conservatives, headed by Chris Ruddy played a key role. Its hourly updates flooded talk radio stations and conservative speakers with the very latest in the party line and in attacks on the liberal left. Every misstep of Kerry or Hillary was instantly sent out to their hundreds of thousands of Internet users and from there went out over the airwaves on talk radio shows from coast to coast. The spinal cord of Hillary's so-called vast right-wing conspiracy, NewsMax's Web page became the sixth most visited Internet news site, fully competitive with those of the leading newspapers, TV stations, and news magazines.[2] Its monthly magazine fired full broadsides against the Democrats on key campaign issues.

Taking their cue from NewsMax, conservative talk radio hosts inundated the radio with daily doses of partisan rhetoric. Rush Limbaugh, who continued to have no peer, led the way. His almost twenty million weekly listeners tuned in religiously to hear the latest from the campaign battlefronts and were never disappointed.

But Sean Hannity, a double-barreled threat with his twelve million radio listeners and three million television viewers, conducted his own political campaign on behalf of the Bush ticket. Touring the country like a candidate, he held rallies in swing states and raised his radio show to a new level of political relevance. On television, he dueled with liberal guests and his Democratic colleague Alan Colmes every night about the key developments that unfolded as the campaign progressed.

Other national radio talk shows also brought their perspective to the debate. Laura Ingraham's iconoclastic style attracted a younger audience often left out by talk radio. Former Nixon staffer G. Gordon Liddy rallied the true believers with his conservative, pro-war show, which specialized in pounding liberals into the dust. Fox News' Bill O'Reilly supplemented the reach of his prime time TV show with a nationally syndicated talk radio show, where he "bloviated" (his term) at greater length on the topics he would cover on television that night. And Fox News anchor Tony Snow added witty and insightful perspective from the platform of his talk show.

But it was local talk show hosts who did the most to polarize and enliven the political environment. Whether it was Neal Boortz in Atlanta or Mancow in Chicago or Mark Larson in San Diego or Al Rantel in Los Angeles or Bob Grant in New York or Steven Gill in Nashville or Keith Larson in Raleigh or Howie Carr in Boston or Ronn Owens in San Francisco, these local hosts did more to spread the word for the Bush or Kerry campaigns than even the national party organizations did.

If the Left outgunned the Right at the bookstores, the radio airwaves were dominated by conservatives. While there is an approximately equal number of hardcore Democrats and Republicans in the United States today—about 30 percent for each side—more than two-thirds of the liberals are young, black, or Hispanic and tend to listen to their own radio outlets rather than participate in political talk radio. The bulk of conservatives are white, and talk radio is their thing.

Movies, too, began to carry the partisan message. Formerly the only form of escapism to remain aloof from the real world, films now dove deeply into the partisan dialogue, both at theaters and in DVD form at video stores. If radio was essentially conservative in its orientation, Michael Moore made sure the film sector of the campaign bolstered the far Left.

His film *Fahrenheit 9/11*—reckless, inaccurate, and irresponsible as it was—made the antiwar movement central to the political agenda of 2004. His distorted and vicious assaults on American soldiers and his highly edited footage gave people the impression that it was George W. Bush, not poor Saddam Hussein, who was terrorizing the people of Iraq. One soldier, who lost both arms for his country, was outraged that a segment of his interview with Moore was so edited as to leave the impression that he was against the war. In fact he was very proud of the part he played, and he came to feel as badly wounded by Moore as by the enemy in Iraq.

Fahrenheit 9/11 became a recruiting film for the only truly antiwar candidate: Howard Dean. (Kerry, Lieberman, Gephardt, and Edwards had all voted for the war resolution.) The Vermont governor's campaign offered a way to become active in opposing Bush and the war. Spurred by the misrepresentations in the Moore film, hundreds of thousands volunteered in his campaign.

Moore's film, which aired in liberal movie theaters throughout the nation, won the Palme d'Or award at the Cannes Film Festival in antiwar France on May 23, 2004. As the Cannes website reports, "the nine members of the jury stood at the Palais entrance applauding the grand winner of this Festival, exchanging final congratulations with the American Director."[3]

Fahrenheit 9/11 must, by any measure, be seen as a landmark change in American politics. While no reliable estimates exist as to how many people saw it, the number may well exceed ten million viewers. That a movie could have such political impact is totally without precedent in presidential elections—but it won't be without imitators in the future.

To answer Moore's film, we participated in a rebuttal entitled *FahrenHYPE 9/11* (available from www.overstock.com). Our answer, banned from movie theaters by the liberal establishment, sold more than 250,000 copies and helped offset at least some of the harm of Moore's broadside. Another film producer, David Bossie, also answered Moore with his film *Celsius 41.11*, which made its way into a hundred and twenty-five movie theaters and had a wide distribution online.[4]

But more than talk radio or films, the key to the expanded participatory politics of 2004 was the Fox News Channel. Soaring above CNN and MSNBC in its ratings, the newest of the cable news stations featured a format of opinion-oriented talk shows that came to embody much of the debate in the 2004 presidential election.[5] Its top-rated *O'Reilly Factor*, featuring the sharp and cantankerous Bill

O'Reilly, came to define a new genre in journalism, a kind of modern-day equivalent to the radio heyday of Walter Winchell. Leading all other entrants in the cable news ratings race, O'Reilly had 3.3 million households tuned to his station as Election Day 2004 approached. *Hannity and Colmes* wasn't far behind, with close to three million viewers. The more liberal shows on CNN—Aaron Brown and Paula Zahn—lagged far behind at under one million viewers, and Chris Matthews's *Hardball* followed even further back.

With Fox's opinion/talk format and the daily flow of campaign coverage, voters were treated to in-depth exploration of each day's developments. But while the establishment media zeroed in on the candidate's statements and his campaign's offerings in their coverage, Fox News gave priority to the events of the campaign—the bloggers' assault on Dan Rather, the Swift Boat attacks on Kerry's war record, the empty Iraqi ammunition storage dump that dominated the last week of the campaign, and finally, the bin Laden video that arrived a few days before the polls opened.

The very nature of the Fox News coverage—stressing opinion rather than footage of the candidate giving a speech—made the environment around the campaign more relevant than it had ever been. With the Left and the Right competing on the Fox News shows, each development in the ongoing battle provided good copy for Fox's offerings that night.

Beyond the organized news and talk outlets, the individual e-mailers and Web bloggers became the core of the participatory politics of 2004. Undisciplined, self-started, and chaotic, these players in presidential politics marched to the beat of their own drummers. Neither campaign could dictate their talking points or even know what they were saying as their messages whizzed over the Internet every day.

If the Left was drowned out on talk radio, it held its own on the Internet. Through websites such as MoveOn.org, developed during the Clinton impeachment nightmare, the Democrats got out their party line to the faithful. MoveOn.org was founded by Silicon Valley "entrepreneurs" Joan Blades and Wes Boyd. As its website explains,

"Neither [Blades nor Boyd] had experience in politics, [but] they shared deep frustration with the partisan warfare in Washington, D.C."[6] The site grew out of an online petition drive to "censure [but not impeach] President Clinton and Move On to Pressing Issues Facing the Nation." The organization moved on to the antiwar issue when Eli Pariser launched an online petition opposing Bush's policies. After getting half a million online signatures, Pariser became the executive director of MoveOnPAC, a post he still holds. MoveOn has enlisted two million activists in all and even ran its own ads online and on television during the 2004 campaign.

The days when a campaign manager or candidate could control his or her own campaign ended with the coming of the blogger. In the modern age of participatory politics, the candidate is more of a franchise: picked up and carried by each of his advocates in their own private campaigns on his behalf. In the new age of politics, you are what your *supporters* say you are, and your campaign is what *they* want it to be. There is no power on earth that can control or even discipline the bloggers and the e-mailers. In a sense, they *are* the campaign. It is their insight and input that reaches more voters than any media outlet—or even more than all of them combined. Their messages are the ones that resonate.

How is it that these voices have amassed such power? Why are their communications treated with more reverence than those of mass media outlets?

Part of the answer is that the relationships that underscore the blogger and e-mail networks often existed long before the campaign. Using the Internet, the online activist contacts his Christmas card mailing list of friends, family, classmates, colleagues at work, and other peers. It is normal for anyone to give more credit to something their friend or relative says than words he or she hears on television.

But there is also the collapse of credibility of the establishment media that animates the success of the online bloggers and e-mailers. Four major news stories in the past two years have flooded the national and international media—but have turned out to be false:

- In September 2003, the BBC reported that Prime Minister Tony Blair, a key Bush ally in the war in Iraq, ordered that his intelligence agency "sex up" its reports on the likelihood that Saddam Hussein had weapons of mass destruction that could easily and instantly reach Great Britain. The intelligence operative who allegedly got the request committed suicide, and a Parliamentary commission found that the story was bogus—but not before Blair was forced on the defensive for months.
- Two months before Election Day 2004, CBS News ran a story criticizing Bush's military service record—based on documents that turned out to be forgeries.
- One week before the election, the *New York Times* reported that ammunition was missing from an Iraqi storage site and implied that it had been removed under the noses of American troops. The allegation turned out to be unproven and most likely untrue.
- In May 2005, *Newsweek* reported that American soldiers guarding and interrogating terror suspects at the Guantanamo Naval Base in Cuba had desecrated the Koran in front of the Camp's Muslim inmates. After anti-American riots tore through the Arab world, killing at least sixteen people, the magazine sheepishly admitted that the story could not be verified and that their government "source" had backed off it. They were forced to issue a retraction.

With a record like that, is it any wonder that people no longer believe what the establishment news organs write and that they have begun to trust the word of people they know firsthand?

In this new world of cyber-roots activism, it seems that one of the functions of all of the media—Fox News, talk radio, NewsMax, and so forth—is to give the bloggers talking points they can send out to their lists in their own private partisan war. In the interconnected world of the Internet, the function of a campaign and a media outlet is to provide material and information to inform the dialogues that rage about them.

In a sense, this development represents a reversion to the campaigns of the late nineteenth century, when the candidate was not the epicenter of his own race for president. As a succession of silent dark horses—men like Rutherford B. Hayes, James A. Garfield, Benjamin Harrison, James G. Blaine, and Samuel J. Tilden—ran for president, their candidacies usually determined in smoke-filled rooms after days of convention balloting, the real action was at the grassroots level. It was there that the campaign was fought in rival party parades, picnics, rallies, and speeches. It was the organization that counted for all; the candidate was barely a guest at his own campaign.

Obviously, the modern candidate is more powerful than his nineteenth-century forebears. But the balance has swung from campaigns that were entirely candidate- and party-centered to ones that stress the grassroots and the spontaneous, simultaneous work of millions of true believers, party activists, and interested bystanders.

In this environment of popular empowerment, the chances of drafting a candidate into a presidential race are more likely than ever before. If the grassroots decide that Condi will be a good candidate and an effective president, they will be able to illustrate the power of their will online just as they did in the 2004 election.

To trace the ebbs and flows of the 2004 contest is to see the power that ordinary men and women wielded over the outcome.

When the 2004 election began, there was no indication that it would be such a landmark year in American politics. George W. Bush faced no challenge on the Republican side; once Al Gore withdrew, most assumed that John Kerry would be the Democratic candidate.

But early in the Democratic process, an upstart challenger from Vermont governor Howard Dean threatened to upset the Kerry bandwagon. Nobody took Dean seriously. Vermont is a small state well known for its ultra-liberalism. (Its only congressman and likely future senator is Bernard Sanders, the only avowedly Socialist member of Congress.) Being governor of Vermont is about as serious a job as being student body president at a Big Ten campus. Dean's decision to sign the first gay civil union bill in the nation seemed to doom any possible presidential prospects. His strong and strident opposition to American intervention in Iraq—and his hesitation in fully backing even our invasion of Afghanistan—appeared to be the final nails in his political coffin.

The major donors to the Democratic Party discounted Dean and viewed the contest for the Democratic nomination as a fight among Kerry, North Carolina senator John Edwards, Connecticut senator Joseph Lieberman, and House minority leader Richard Gephardt. Dean seemed to have no support.

But Dean's political consultant, Joe Trippi, realized that he could reach hundreds of thousands of antiwar and pro-gay activists and fuel his client's presidential bid through the Internet. At first, Dean's prospects must have seemed remote. As Trippi notes in his book *The Revolution Will Not Be Televised: Democracy, the Internet, and the Overthrow of Everything,* at first the campaign had "seven people on staff, $100,000 in the bank, and four hundred and thirty-two supporters."[7]

But they used the Internet in a way it had never been tapped before; the Dean campaign entered chat rooms, posted on Internet bulletin boards, sent out e-mails, and networked through the antiwar and gay communities. By the end of 2003, Dean was on track to raise $50 million; he raked in $15.8 million in the last three months of the year alone. Trippi writes that Dean had six hundred thousand "fired up supporters . . . people who [had] never been politically involved before and who are now living and breathing [the Dean] campaign."[8]

In 1999, we had written our own book on the coming Internet political revolution, with a similarly lengthy title: *Vote.com: How Big-Money Lobbyists and the Media Are Losing Their Influence, and the Internet Is Giving Power to the People,* which foretold this transition from television to Internet—but it was Trippi, one of the great political consultants of our era, who was actually the first to make it happen.

One of the neat things about Internet campaigning is that it goes on beneath the radar. The establishment news organs had no idea what was happening at the Dean headquarters and continued to treat him as a marginal candidate. So the Kerry campaign, blissfully unaware that their candidate was facing a monstrous challenge on his left flank from a bona fide antiwar candidate, was busy moving their candidate to the center to meet the challenge from his more moderate opponents: Edwards, Lieberman, and Gephardt. Caught by surprise by a challenge on his left from Dean, the Kerry campaign was slow to realize how potent the Vermont governor's operation had become. "There are no votes on the Internet," said Kerry campaign manager Jim Jordan in one of those prophetic remarks made just a few months before he was fired for failing to respond adequately to the Dean challenge.[9]

Energized by a massive influx of volunteers and cash, Dean's campaign was on a roll. When the Vermonter reported having raised more cash in the third quarter of 2003 than any other Democrat, the media had to begin to take him seriously. Afire, the Dean campaign "had burst from the pack with an Internet-fueled campaign," *Newsweek* reported.[10]

Before the Internet, a candidate could fund his campaign in only three ways: He could put up his own money, rely on big donors and special interests, or try to amass funds through direct mailings and telephone solicitations. Dean had no money of his own, and the party establishment wasn't about to start funding such a liberal long shot.

But direct mail or telephone fund-raising weren't the answers either. To solicit money by mail, a campaign must lay out massive amounts of cash to send prospecting letters to millions of possible donors. Typically, these mailings cost at least fifty cents each to produce and mail, even at bulk rates. The campaign usually needs to get 2 percent of the people it solicits to send an average donation of twenty dollars just to break even and cover the cost of the mailing. A return this high is very, very hard to get and almost impossible to predict. Yet a campaign finance director has to devote millions of up-front dollars trying to reach these levels. He sends out the mail, tying up all the campaign's cash, and crosses his fingers that it will work.

But to survive, of course, a campaign has to do more than break even. The purpose of the mailing is to raise money for a much bigger event: the television campaign. Direct mail almost never makes money in its initial outreach. For that, direct mail campaigns depend on resoliciting the unfortunates who send in their money in response to the initial mailing. These unsuspecting victims have now earned the right to be dunned, about every six weeks, for more and more money. The campaign then hopes for a 10 percent response each time from these proven donors so that the real money will begin to flow in.

Telephone solicitation is even more problematic than mailings and often has higher unit costs. It costs up to a dollar to pay an allegedly live human being to call each voter and read him or her a scripted pitch. A tape recording is cheaper, but much less effective. And the ultimate success of any such venture depends on the quality of the lists the campaign is mailing or phoning. If the list is too narrowly drawn, it won't raise enough money. If it's too widely constructed, it will cost too much money in its initial mailing.

But the Internet has changed the whole calculation of grassroots fund-raising. Since there are no postage or telephone charges online, a political campaign can e-mail frequently and massively to possible donors, unconcerned with coming up with initial seed money to fund the outreach. Direct mail responses have to be opened, tallied, and deposited by hand, whereas Internet donations are usually given through credit cards and can be used immediately.

And while it takes weeks to send out a fund-raising solicitation in the normal post—"snail mail"—and get a response, it took the mere click of a button to resolicit the Internet donors who formed the core of the Dean political base. This capacity to reload quickly proved essential to the momentum Dean was able to achieve.

Dean raised tens of millions under the guidance of Trippi, and soon he was mounting a serious challenge to Kerry. The upheavals of 2004 had begun. A candidate had come from nowhere to challenge the party's front-runner, paying his way through grassroots fund-raising. Where Kerry and the other Democratic candidates could boast tens of thousands of individual donors to their campaigns, Dean could point to hundreds of thousands—all because of Trippi's Internet savvy.

But Trippi and Dean did more than just use the Internet to raise money for television commercials. Online solicitation can raise money, of course, but using it exclusively to generate funds is like using the Air Force merely to ferry troops into combat. The Internet campaign Trippi waged built an entire community of active volunteers, eager to help Dean get elected. As the campaign sent out newsletters and reports of Dean's activities and schedule, people felt that they were more than just names on a fund-raising list; they belonged to a group that was trying to change the country. This sense of joining animated the entire Dean campaign and generated an enthusiasm that outlasted his electoral efforts. In 2005, when Dean sought successfully to keep his political career alive by becoming chairman of the DNC, his Internet organization was alive and well and eager to help him in his new quest.

As Dean gained momentum in 2003, Bill and Hillary Clinton, initially closet Kerry supporters, began to fear that the Vermont governor, who bore them no love, would win the nomination. In a panic, they encouraged General Wesley Clark to enter the race. The Clinton operatives, who had been working for Kerry, led by the Massachusetts senator's campaign manager, Chris Lehane, switched to Clark.

As the Democratic establishment came to fear a Dean nomination, convinced that he would lose massively to Bush in November, they began to pelt the liberal candidate with negative stories. Suddenly Dean was on the firing line for refusing to release his gubernatorial papers and for offhand comments on the campaign trail. Under pressure and hounded, Dean made a massive mistake and fired Trippi.

Eventually the combined pressure of the Democratic Party establishment—and its journalistic lackeys—proved too much for Howard Dean's campaign; he faltered, first in Iowa and then in New Hampshire, and fell out of the race.

The establishment had won its first encounter of the 2004 election season with the forces of grassroots political action. But it was a close call and one that revealed the power of the previously docile electorate. It took all the king's horses and all the king's men to put the John Kerry campaign back together again after it was challenged by the cyber roots that underlay the Democratic Party base. Previous candidates had come out of nowhere to challenge the front-runners in their party's presidential nominating process. Bill Bradley and John McCain mounted highly credible and unexpectedly strong races against Gore and Bush in the primary season of 2000. Gary Hart came close to upending former vice president Walter Mondale in 1984. Further back in time, neither Barry Goldwater nor George McGovern had been front-runners before when they won their party's presidential nominations in 1964 and 1972.

But 2004 marked the debut of the Internet as the fuel behind an upstart candidacy. After all, Dean never won a primary or a caucus. It was his success at raising money online that made him the front-runner before a single ballot was cast—all because of the activism of his Internet ranks.

And, in a larger sense, Dean proved that it was easier, cheaper, more productive, and faster to raise funds from a mass base of Internet donors than from a few rich donors and favored special-interest PACs. For all the efforts of campaign finance reformers like John McCain and Connecticut congressman Chris Shays, it is the mobilization of Internet donors that is hastening the end of special-interest domination of campaign funding. It's a revolution few expected to see in our lifetime. Clean money has become more accessible than dirty special-interest dollars. No longer must a candidate sell a piece of his soul with each fund-raising event. If he seeks funds online from those whose only connection to him is that they agree with him on the issues and admire his candidacy, he will not only be cleaner, but richer too!

To defeat Howard Dean and regain the momentum his campaign had lost, John Kerry went back in time—back before his candidacy, before he was a senator, before he was lieutenant governor of Massachusetts, back to the time when he was just a lieutenant in Vietnam. Dennis Rasmussen, a member of his old Swift Boat squadron, joined Kerry on the campaign trail. His moving story of how the candidate had saved his life at risk to his own mesmerized Iowa voters. The Democrats, sensitized to the need to show toughness in the War on Terror and traumatized by having to defend Bill Clinton's draft dodging, were thrilled to embrace their own version of a war hero.

Kerry needed the Iowa win to have a chance in the first real primary state: New Hampshire. He and Dean both came from neighboring states, so it was a must-win for each.

It was the genius of Kerry's campaign to decide to contest Iowa as a way of winning New Hampshire. In the remote Midwestern farm state, Kerry figured he had a better chance to derail Dean than in the New England base they shared. As soon as New Hampshire Democrats read the Iowa returns, they switched in massive numbers from Dean to Kerry.

New Hampshire voters, long the epicenter of presidential primary politics, were accustomed to voting tactically. They saw their primary as a kind of semi-final, where they got to choose the two candidates who would then go head throughout the rest of the nation competing for the nomination. They felt it was their function to clear the field of the also-rans.

Liberals—an outnumbered species in a New Hampshire general election, but the controlling faction in a Democratic primary—especially felt an obligation to make sure one of their soulmates survived to fight for the nomination in the larger states with later primaries. Initially they had bet on Dean, but now that Iowa had spoken, they came to doubt his staying power. They needed a strong leftist candidate to challenge the likely semi-finalist from the center, John Edwards, so they switched to the next available liberal and gave the primary to John Kerry.

After New Hampshire, the contest was basically over. Edwards couldn't stop the Kerry momentum; he surrendered a few weeks later.

In the course of the primaries, though, Kerry's biography became a substitute for strategy as the candidate, his wife, and his handlers scrambled to deal with the foremost issue in the campaign: the war in Iraq. Desperate for some kind of bounce at their convention, the Kerry team decided to feature his war record front and center during the candidate's acceptance speech—usually the seminal event in a campaign.

Surrounded by his Swift Boat comrades, Kerry mounted the platform in Boston to speak to the convention and the entire nation on television. He saluted and began his speech by saying "Lieutenant Kerry reporting for duty." The move turned out to be a huge blunder and set the stage for Round II of the real battle of 2004: the establishment media vs. the bloggers and the New Media.

Americans did not want a lieutenant as their chief executive. They wanted a commander-in-chief. Had Kerry chosen to speak about his Senate career, or even his adulthood, he might have gained traction with the voters. By dwelling on his days in Vietnam, however, he fell far short. He came out of the convention with virtually no bounce in the polls.[11]

But that wasn't half the story. It soon turned out that his record in Vietnam was not all that he said it was. And now the Internet, and the new participatory politics, would strike again.

Watching the Democratic candidate extol his war record were a group of men who were not quite so enamored of Kerry's military valor. Swift Boat veterans themselves, they felt that Kerry was misrepresenting his war record. They didn't like that his three Purple Hearts entitled him to leave Vietnam before they did. To make matters worse, they suspected that the final one was undeserved. And once Kerry returned to the states, they disliked how he dumped on them and implied that his old comrades in arms were war criminals.

But the vets had no money—or not much, anyway. All they managed to scrape together was two hundred thousand dollars from a Texan named Bob Perry. Undaunted, they set out to run some television ads in a few markets around the country challenging Kerry's depiction of his war record. Armed with the modern campaign equivalent of a slingshot, they decided to try to change the world. But they were realistic. According to *Newsweek,* "the Swift Boat vets didn't think they'd cause much of a stir . . . they were pretty sure the establishment press would just blow them off."

They were right and they were wrong. The establishment press did just blow them off—totally. But then, after the networks and the leading newspapers in the nation sloughed off their attacks, the alternative media—the Internet, talk radio, cable television, and private bloggers—picked up their story and upended the entire Kerry campaign.

The first ad the Swift Boat vets bought with their slender cash reserve ran in parts of West Virginia and Ohio (two swing states) and accused Kerry of lying to get his medals.[12] The second charged that when he returned to the States he turned on his former mates and joined actress Jane Fonda in her antiwar crusade. The ad showed a young, hairy John Kerry testifying before Congress, accusing U.S. troops of war crimes.[13]

Kerry's advisors missed the point again, just as they did when they were blindsided by the surge of the Dean campaign. They pooh-poohed the Swift Boat ads, failing at first to see how damaging they could be.[14] After all, the Swift Boat folks could only afford a minuscule advertising budget; they had to cancel a press conference in Washington, D.C., due to a lack of media response.

But the ads began to swim through the arteries and veins of the new alternative media. Bloggers alerted one another, talk radio shows reported the accusations about Kerry's record, and Rush Limbaugh and Sean Hannity brought the Swift Boat vet message to tens of millions of listeners. NewsMax tipped off local talk radio hosts throughout the nation. On Fox News, *The O'Reilly Factor* and *Hannity and Colmes* began to cover the story, and the charge began to stick: Kerry's medals were a fraud.

The story spread underground and online. Bloggers joined the act, spreading the word to tens of millions of voters who were well beyond the reach of the meager financial resources of the Swift Boat vets themselves.

The Kerry campaign compounded the damage by failing to answer the charges, feeling that to do so would just call more attention to them. Kerry advisors Bob Shrum and Mary Beth Cahill argued that the ads only appealed to the Republican base, so what was the point of answering? When the networks, the *New York Times,* and the other organs of the journalistic establishment failed to cover the story, the Kerry people felt even more confident in their decision not to respond. But it was a classic mistake.

The Swift Boat charges began to bite into Kerry's image. The candidate had set himself up for their charges by banking everything on his war record. In retrospect, one wonders how Kerry could possibly have staked his entire pursuit of the presidency on what happened in the Mekong Delta thirty-five years ago.

It wasn't so much that people began to believe Kerry's war record was suspect. Few really cared if he had been shot twice or three times. But voters were very concerned about the idea that Kerry might have lied about his record. Bill Clinton's misadventures with the truth were still fresh in everybody's mind. Voters blamed themselves for not spotting Bill's flaws during his 1992 campaign and giving him a pass on the accusations that arose back then. They were determined not to make the same mistake again.

The debate persisted in the Kerry camp on whether to answer the charges or remain silent.[15] Kerry himself wanted to respond in a speech to the Veterans of Foreign Wars on August 18, 2004, but his campaign convinced him not to. The vice presidential candidate, John Edwards, wanted to speak out as well, but Stephanie Cutler, Kerry's communications director, vetoed it. Why call attention to the Swift Boat accusations?

Blind to the new alternative media's potential to carry a story around the country, the Kerry operatives felt that if it wasn't being carried in the *New York Times* or the *Washington Post,* or on CBS, NBC, or ABC, or in *Time* or *Newsweek,* it wasn't happening. If the vets were only spending a quarter of a million dollars, how could the ads be having any effect? For decades, political veterans had been discounting just this kind of loosely organized, ill-funded attack, successfully burying it with their own positive ads and selected newspaper rebuttals. But this time it didn't work. Because now the Internet and direct one-on-one communication through e-mail and blogging had lent power and momentum to the story and the Swift Boat accusations.

The story flooded the country under the noses of Kerry's blind advisors. Unwilling to take the Swift Boat vets on directly, the Kerry campaign finally began to feed stories and documents to the mainstream organs like the *New York Times,* the *Boston Globe,* and the *Washington Post* to challenge the accusations. Surely, they reasoned, once these mighty outlets had spoken, the charges would lose their credibility. But they did not, and the establishment never caught up with the alternative media and the Internet.

By the time of the Republican National Convention at the end of August, Kerry's credibility was in tatters and his efforts to base his campaign on his war record had backfired massively. For the first time, citizens, acting on their own, had been able to defeat the vast public relations machine of the Democratic Party and the national media attention that had been focused on Kerry's speech at his convention. The bloggers had won Round II.

But the liberal media establishment had a backup plan. The left-leaning CBS News anchor Dan Rather and his *60 Minutes II* crew were planning a major post-Republican convention exposé on President Bush's stint in the Texas National Guard. CBS claimed that they had records, which had never been disclosed, from the files of Bush's commanding officer indicating that he had often failed to report for duty and had used political favoritism to avoid punishment or discipline. If true, the charges threatened to offset the attacks on Kerry and bury Bush in the minutiae of whether he reported as ordered thirty years ago. As *Newsweek* put it in its campaign wrap-up, "For a moment it looked as if the tables had turned, and Bush would have to endure an uncomfortable round of questions about his spotty attendance record in the stateside guard while Kerry had been dodging bullets in the Mekong Delta."[16]

60 Minutes and *60 Minutes II* are among the top-rated news shows on American television. Drawing more than twice as many viewers as the network's nightly news shows, they carved out a niche for investigative reporting that compelled America's attention. After all, it was *60 Minutes* that had given Bill and Hillary a chance to rebut the charges of Gennifer Flowers during the 1992 campaign. "We saved their ass," Don Hewitt, the program's creator, told me.[17]

For *60 Minutes II* to air such an accusation against a sitting president within two months of an election was a heavy hit from the establishment's biggest gun. But the attack had an Achilles heel: It rested on "documentary" evidence, a handful of letters from Bush's commanding officer setting forth his transgressions and how political intervention had covered them up. And the letters turned out to be forgeries.

The first one to criticize the documents on which the story was based was first lady Laura Bush, who said that she doubted their authenticity.[18] Soon others joined in. Internet bloggers throughout the land started scrutinizing the text of the documents Rather had relied on, which were supposed to have been written back in the precomputer era. But the "th" listed after dates was raised (as in 14th), a feat that would have been impossible with a regular electric or manual typewriter back in those Cro-Magnon years.

Word of the forgery raced through the talk radio circuit and the same venues that had publicized the Swift Boat story, discrediting the CBS documents and the allegations that flowed from them. The mainstream media, blind as always, initially refused to deign to cover the attacks on the *60 Minutes II* story, and CBS refused to concede an inch.

But the attack was relentless, and soon its focus included not only the network's credibility but the Kerry campaign as well. Apparently the story's producer, Mary Mapes of *60 Minutes II,* had called Kerry aide and former Clinton press secretary Joe Lockhart to offer to put him in touch with her source for the forged letter, retired Texas National Guard lieutenant colonel Bill Burkett. In the middle of a political campaign, it was a call that even the executives at CBS recognized as a conflict of interest.

Finally, long after the damage to CBS, Dan Rather, and John Kerry had proven irreparable, the network backtracked and admitted error. The longtime newsman "voluntarily" retired as anchor, his reputation ruined by his clumsy attempt to tilt an election to the Democrats.

By mid-September, some polls showed Bush ahead of Kerry by as much as thirteen points. The election seemed to be over. The one-two punch of the bloggers' successful destruction of Kerry's image and the rebuttal of attacks on Bush's National Guard record had inflicted maximum damage on the Kerry campaign. The bloggers had won Round III.

But the fundamental reason for Bush's advantage over Kerry wasn't just the tactical edge he had in the exchange over these two issues. The Bush people understood—as the Kerry advisors never did—that the president was permanently ahead of his Democratic rival on the issues of terrorism, national security, and defense. When voters were asked whom they trusted more on these topics, they overwhelmingly answered Bush, even if they subsequently voted for Kerry in the poll. On the president's worst day in Iraq, when dead American servicemen and women were piled high, voters still trusted their president more on these issues. After all, Bush had earned their trust through his resolute stand in the aftermath of 9/11 and in his successful invasions of Afghanistan and, to a point, Iraq.

But on domestic issues like social security, health care, the environment, and education, John Kerry had a similarly strong advantage. Voters trusted the Democratic agenda on these issues much more than they did the GOP approach. For seventy years, the Democrats had led the way on these questions while the Republican Party dragged its feet—or dug in its heels—and Kerry was in a position to benefit. On the economy, voters generally divided evenly between the two candidates, so it didn't do much to tip the election one way or the other.

So the Bush people understood that there was no such thing as a bad day in Iraq or in the War on Terror. As long as these topics dominated the media, their man had a huge advantage. But when the topic of discussion turned to virtually any aspect of domestic policy, the Democrat would have the edge. This fundamental strategic construct dominated all the GOP thinking about the presidential race: It wasn't what was said. It was the topic that mattered.

The Bush campaign had a clear strategy: Talk about terror as much as possible and about everything else as little as you can. Its day-to-day tactical decisions were all in pursuit of this basic strategy.

But the Kerry campaign had no strategy, just a series of tactics. Each day the campaign would grope for an edge, whether they found it on the domestic or the foreign front. Like many political advisors and consultants, Kerry's people overestimated their ability to impact public opinion. There was really nothing Kerry could have done to persuade voters that he was the better candidate for commander-in-chief, and all he accomplished by trying to do so was to heighten the salience of these issues. Similarly, no words George Bush could utter would convince voters that he was better on issues like health care and social security.

Karl Rove and the Bush strategists understood this fundamental fact. But the Kerry advisors did not. They clung to their belief that their candidate could win on the Iraq issue by focusing on the seemingly endless casualty lists and on the War on Terror by pointing out that bin Laden was still alive and at large.

Bush was playing chess—focusing on a strategic vision. The Kerry campaign, on the other hand, was playing checkers—looking for advantage in each day's news cycle, without any overall strategic plan for victory.

What were Kerry's staff and consultants doing while their candidate was getting fried? Their arrogance, self-promotion, and narcissism seem legendary, even by the diminished standards of the political consulting profession:

- According to *Newsweek,* late one Saturday night while he was boning up for a TV appearance the next morning, Kerry called his advisor Tad Devine, the partner of political consultant Bob Shrum. Devine admonished him not to call so late. Devine told his candidate, who was staying up late studying his briefing papers, "The most important thing is that I get a good night's sleep. Don't call me."[19]
- On the Saturday of Labor Day weekend, James Carville, Clinton's 1992 campaign manager, threatened Kerry manager Mary Beth Cahill that unless she gave "effective control [of the Kerry campaign] to Clinton's former press secretary Joe Lockhart," he would "go on *Meet the Press* the next day and 'tell the truth about how bad [the Kerry campaign] is.'" Carville's threat worked. Nobody dared to defy the Ragin' Cajun and the campaign endured a "silent coup."[20]
- After Bill Clinton called Kerry from his hospital bed and counseled him to spend less time talking about Vietnam and more time fighting Bush over Iraq, word of the conversation leaked.[21] The story appeared on the front page of the *New York Times,* making Kerry look like a fool who was unable to run his own campaign. Some Kerry campaign aides blamed Lockhart; others said the leak came from Carville. I think it may have come from Clinton himself. But one thing is certain: It didn't come from Kerry. One wonders if Carville and Clinton really wanted Kerry to win at all. After all, had he prevailed in 2004, Hillary Clinton would have no shot at the White House in 2008. In 2012, would-have-been vice president John Edwards would probably have gotten the Democratic nomination—potentially putting Hillary out of the running until 2020, by which time she will be seventy-three years old.

As the time approached for the presidential debates, Bush seemed to enjoy a big advantage: The first, and most widely watched, match would be exclusively on foreign policy, the president's home turf. But Bush was decidedly off that day. His responses were vague and rambling. He seemed peevish and petulant. He smirked when he should not have. Bush seemed almost muscle-bound with overpreparation and unable to handle issues with even his normal below-average verbal dexterity. Incredibly enough, Kerry decisively won the foreign-policy debate, a battle in which Bush should have easily prevailed.

The president righted himself in the second debate, but the impression his first performance had left was hard to erase. And in the vice presidential debate, Dick Cheney wiped the floor with the Democratic ingénue running against him. As Cheney corrected Edwards' factual misrepresentations again and again, he seemed like a university professor reviewing the paper of a freshman student: "You got this wrong and that. Half credit for this. Next time, son, do your homework." But vice presidential debates do little to sway voters.

Under the ground rules negotiated by Bush and Kerry's advisors, the third and final debate between the presidential candidates would be exclusively about domestic issues, Kerry's territory. For once, Kerry's people had outsmarted Bush's. The Republican negotiators should have resisted designating the subject matter for any of the debates. Without any predetermined subject, terrorism would naturally have dominated all three debates. After all, how can you interrupt an answer about terrorism to discuss the fine points of academic standards in elementary schools?

But the deal stood, and Kerry faced the prospect of entering the final two weeks of the election with a good head of steam after a debate centered exclusively on his area of strength. The president, stunned by his earlier debate defeats, doing his best to be dynamic and aggressive, scored many points, but the focus of the last debate—on education, health care, Medicare, and social security—was so skewed in Kerry's favor that there was little he could do to prevent the Democrat from winning.

Kerry, who had seemed dead as he entered the debates, acquired a new life. After the debates the race was nip and tuck, with no clear advantage for either candidate. And the end game looked bleak for Bush. The domestic subject matter of the third debate seemed to guarantee that, in the final two weeks, voter attention would be riveted more on the homegrown issues that favored Kerry than on international affairs.

Kerry could have won. But once again his people's incompetence, the Bush campaign's skill, the overreaching of the establishment media, and the vast network of bloggers and Internet e-mailers combined to deliver the win to the incumbent president.

Round IV of the establishment media vs. the bloggers opened on a Monday, eight days before Election Day, when the *New York Times* broke a story it had been developing with *60 Minutes*. The *Times* reported that 380 tons of explosives had disappeared from an ammunition dump in Iraq after U.S. troops had moved in to secure the outpost. The implication of the story was that the administration had jeopardized American lives by failing to secure the site and had given the terrorists precisely the weapons they needed to kill our troops.

60 Minutes, partisan to the end, had planned to air the story on the Sunday night before the election so as to guarantee that Bush could not recover. With only forty-eight hours left before polls opened, the Democrats at CBS probably reasoned that they could hit the Bush campaign so late that it couldn't answer. But the *Times,* perhaps in a fit of responsibility, decided to go with the story earlier.

The accusation was based on a letter written by Hans Blix, the chief UN weapons inspector, indicating that the explosives were present when the final UN inspections had taken place right before the start of the war. The site was now empty; the *Times* hypothesized that the Iraqi terrorists had looted it after the American troops had taken over.

Kerry's campaign jumped right into the fray, canceling its regularly scheduled television ads to hone in on the new charge. Kerry assailed Bush for negligence and called the president "incompetent." He bet the election that he could convince the public that the administration had blundered. This candidate, who staked his convention speech on the murky events in the Vietnam jungle thirty-five years ago, was now tying his chances for victory to something that happened at a desert ammunition storage facility during the confused and chaotic days after the invasion of Iraq.

The Pentagon raced to get to the bottom of the story, eventually issuing documents and photos that suggested that the dump had most likely been emptied by American troops and the ammunition was safely in allied hands. But it was one thing for the Pentagon to issue explanations and quite another to assure that the answers would catch up with the initial charges. As Mark Twain once said, "A lie can travel halfway around the world while the truth is putting on its shoes."[22]

And the establishment press wasn't helping Bush get his rebuttal across. No matter how specific and convincing the Pentagon's explanations and evidence, the *New York Times* maintained its skepticism, and its attitude permeated the coverage in newspapers throughout the country. The Pentagon evidence was subjected to unfounded speculation and contradicted by vague theories on front pages throughout the nation.

In the meantime, Kerry pumped up his television campaign, running ads that echoed the *Times* story. In the past, this alignment of the establishment mainstream media and a paid advertising buy of enormous proportions would have spelled death for the Bush campaign. The president would have had to take his own ads off the air and debate the veracity of the *Times* charges through an ad war—preempting any attempts to get his own message out in the crucial last week of the campaign.

But in 2004, the bloggers, Fox News, and talk radio all highlighted the Defense Department's swift and effective rebuttal. They went over every detail of the Pentagon's evidence and ultimately discredited the charges against Bush. Even though the *New York Times* and other establishment news organs still treated the matter as if the administration may have blundered, the speed and reach of the alternative media far outweighed any advantage the establishment may once have had. It soon became evident that the American people weren't buying the charge. Round IV went to the alternate media and the bloggers.

Meanwhile, Kerry had managed to snatch defeat from the jaws of victory by refocusing the entire election on who could do the best job of winning the war in Iraq. By jumping on the *Times*'s allegation and focusing on it in his stump speeches and television ads, he brought the spotlight back to terrorism, Iraq, and foreign policy, Bush's areas of strength. Unaware of the strategy Bush's people were pursuing, Kerry pounced on what he thought was a tactical advantage afforded by the *Times* story—to the detriment of what should have been his strategic game plan.

All the focus on health care, the economy, social security, the environment, Medicare, prescription drugs for the elderly and education—Kerry's issues—was lost. Now, suddenly, the election had boiled down to a simple question: Who would make a better commander-in-chief: the sitting president of the United States or the former lieutenant-turned-senator running against him? It was a fight Kerry couldn't hope to win.

Finally, on the weekend before the election, Osama bin Laden weighed in with a videotape in which he seemed to threaten massive terror against America and a repeat of "Manhattan" [that is, the 9/11 attacks] if the elections produced a victor who would oppress the Islamic people. The terrorist impudently put words in the mouths of the 9/11 victims, claiming that it was as if they were telling the American people to "hold to account those who have caused us to be killed."[23] After savaging Bush repeatedly, he said that "an ounce of prevention is better than a pound of cure" and warned Americans that "your security is not in the hands of Kerry, nor Bush, nor al Qaeda. No. Your security is in your own hands." The implication was clear: By rejecting "the liar in the White House," America could buy its way out of further attacks.

What was Kerry to do with bin Laden seeming to endorse a vote against Bush? Rather than keep quiet about it—or condemn bin Laden for intervening in a U.S. election—the Kerry campaign used the tape to bewail the administration's failure to catch the terrorist. But to some it seemed a confession of bin Laden's impotence that he had to send a tape to America. After all, when he wanted to tilt the Spanish election, he had bombed a train instead.

In the days before the election, according to GOP chairman Ed Gillespie, the Bush campaign mobilized a massive force—1.6 million volunteers—to work in swing states.[24] Days or even weeks before the election, these unpaid workers traveled to key districts, getting to know the voters and establishing relationships with them. When the polls opened on Election Day, they returned to visit those they had canvassed, reminding them to vote. Using an unprecedented mass of volunteers, Bush managed to draw to the polls twelve million more supporters than he had in 2000. Kerry, relying on fewer workers, most of whom were paid, had a far less intense field operation. It was a matter of making a virtue of necessity. Democrats traditionally relied on labor unions to get voters to the polls; this time, Republicans were aware that they couldn't count on similar support, and took on the task themselves and recruited their own people.

One reason for the Kerry campaign's lack of good on-the-ground coverage was a fake-out by the Bush campaign. Republican leaders understood how alarmed Democrats and the Gore campaign had been by the reports of massive voter fraud during the 2000 election. These allegations, many of them bogus, suggested that Republicans had done their best to suppress black turnout, intimidating tens of thousands into not voting.

Playing on Democratic paranoia, some Republican operative had the clever idea of spreading the word, before Election Day, that the Republican Party was planning to make tens or even hundreds of thousands of challenges to voters as they tried to cast their ballots. In Ohio, the rumor spread that the Republicans had a long list of names of people they claimed were wrongly on the voting rolls. They were thought to be equipping GOP poll watchers with the information for use on Election Day.

Democrats complained feverishly that such harassing tactics were really designed to lengthen the lines at these voting booths, discouraging voters in Democratic districts from waiting to vote. Even if the challenges failed, the Democrats said, they would help Bush by decreasing the Kerry vote.

To counter the rumored GOP effort, the Democrats mobilized ten thousand lawyers and trained them in the arcane arts of Election Day voting rules. So absorbed was the Kerry campaign in defeating the expected Republican voter challenges that it never put the time and planning into Election Day field operations that was required.

On Election Day, it all turned out to be what Gillespie called "a head fake."[25] The Republicans didn't challenge voters as they arrived at the polls, leaving the Democratic lawyers with little to do. In the meantime, the Republicans were able to outgun the Democrats in the field operations on which they had really been concentrating all along.

The results were not only a victory for Bush and the Republicans, but a win for the participatory politics of the future. From the Swift Boat vets to the bloggers, Fox News hosts, and the rest of the alternative media, it was a grassroots groundswell that made all the difference. They had defeated the best the establishment had to offer—the *New York Times* and CBS—and, in the process, changed politics forever.

From now on, politics will be dominated from the bottom up. The elites have been outgunned by ordinary people. The tail has stopped wagging the dog. Mass media has been transformed from a top-down process to one that forces the media elite to listen to what the masses are saying—or risk drifting into irrelevance. And in this new age the Republicans have the advantage—from their higher level of Internet use to the greater literacy of their voter base. In 2004, the wealthy and highly educated Republican base—with its greater ability to devote free time to politics—overwhelmed the best efforts of the less educated, less galvanized Democrats. And those odds are not likely to change anytime soon.

It is in this environment of bottom-driven politics that we contemplate the possibility of a genuine grassroots draft movement to encourage Condoleezza Rice to run for president.

In the new political world, for example, it is possible to raise large amounts of money without forcing a candidate to press the flesh endlessly, making deals with special interests and massaging fat cat donors. If a candidate has a strong appeal to a large number of people, as Condi certainly does, her supporters can reach out to one another and raise the funds for a race without Rice herself ever picking up the phone.

In the past, when funds had to come from large donors or special-interest PACs, such indirect fund-raising was impossible. No rich donor will give money without being stroked by the candidate. Whether she has to schmooze with him at a cocktail reception or take photos with him to boost his ego, the art of extracting money from a campaign donor requires a lot of hands-on time from the candidate herself.

But with the Internet, the idea of Condi for president will attract millions of supporters around the nation. All that is needed is to harvest the money through aggressive online networking. The candidate is no longer central to the fund-raising process.

In the old political era, one had to depend on the party machinery in each state to get the word out to voters about one's candidate. With most voters staying home and watching politics on television, it was left to the regular party activists to work in campaigns for presidential candidates. But in the era of grassroots participation, new volunteers will flock to the "Rice for President" campaign, providing an instant organization in each of the fifty states.

While the other candidates need to assemble their organizations laboriously, brick by brick, Rice's reputation will have so preceded her that her organization will sprout as soon as word gets out that there is a movement to draft her into the presidential contest.

How will the mainstream media react to a draft-Condi movement? As usual, they'll probably miss the point until it's too late. They will focus on the field of active candidates, each vying for their moment in the spotlight, and will miss the grassroots groundswell of support for the noncandidate. They will overlook the fact that, in this era of participatory politics, the real action is happening down below, not up above. Just as they never registered that the Swift Boat vets could turn the campaign on its head or that the bloggers could defeat Dan Rather, they will be slow to realize how a large number of aroused voters can create a presidential campaign without the media, the manipulators, and the money that usually surround presidential campaigns.

But Hillary and Condi are not the only games in town. They have plenty of company in the presidential field. Might Hillary be derailed before the convention? Will another Republican step forward who can defeat her, making a Condi draft unnecessary?

11

Who Else Is There?

In 2008, one of those rare occasions will occur when both parties have the chance to make an open choice of their nominee, without either having an incumbent president certain to be renominated. Since World War II, the stars have aligned this way only five times: in 1952, 1960, 1968, 1988, and 2000.

In each party, possible contenders are already lining up at the starting gate. There are so many—at this stage in a presidential contest, optimism knows no bounds—that there would seem to be little room for a draft, even for someone as compelling as Condoleezza Rice. But a close examination of the field shows otherwise.

Let's start our survey of the contenders by examining what is the fundamental premise of this book: that Hillary Rodham Clinton is the certain nominee of the Democratic Party.

THE DEMOCRATS

At the moment, six potential contenders for the Democratic nomination have been mentioned with any seriousness: Hillary, John Kerry,

John Edwards, Howard Dean, Delaware senator Joe Biden, and Indiana senator Evan Bayh. More are likely to join the field. Even without a realistic hope of winning, the chance to join the early candidate debates—when eight or ten candidates line up as if at an auction—is irresistible to some attention-starved politicians.

As noted, the polls show Hillary far ahead of the pack, with twice as much support as Kerry, and three times as much as Edwards. But early polls can be deceiving.

Kerry and Edwards, of course, share one big problem: They came out of the 2004 election as damaged goods.

Ever since Adlai Stevenson lost his second consecutive race for the presidency in 1956, no candidate has ever been renominated in the next race after a defeat. (Richard Nixon lost in 1960 and won in 1968, but he sat out the 1964 election precisely to regain his stride and recover his support.) Indeed, in the entire history of the country, only five candidates—Charles Pinckney (1804 and 1808), Henry Clay (1824, 1832, and 1844) William Jennings Bryan (1896, 1900, and 1908), Thomas E. Dewey (1944 and 1948), and Stevenson (1952 and 1956)—have ever lost more than one election as the nominee of their party.

Losing an election, after all, is a difficult experience for both a candidate and his supporters—and no defeated standard-bearer comes out unscathed. While Hillary's popularity may account for some of her margin over Kerry in the polls for the Democratic nomination in 2008, Kerry burned many of his bridges with party activists and loyalists with his dismal performance in 2004.

Kerry, for his part, doesn't seem to realize that he lost the election. Nearly six months after the election, he was quoted by *Time* as claiming that "we actually won in the battleground states."[1] Oh, really? What about Ohio, West Virginia, Kentucky, Missouri, Florida, New Mexico, Nevada, and Iowa, all of which went for Bush?

Kerry blames his defeat on the fact that the Republican team had "six years to develop its electoral strategy," while he had only eight months. *Time* suggests that the Massachusetts Democrat sees November 2, 2004, as "merely a detour on his road to the White House" and reports that he was planning to rack up frequent flyer miles traveling to twenty cities in two months to further his 2008 chances.[2]

Incredible as it sounds, Kerry has $14 million left over from his defeat. (What does it say about a candidate that he would lose an election while still holding a reserve of funds in the bank?) Kerry is now handing out his money to other Democrats, trying to curry favor for his next presidential bid.

But realists are trying to break the news gently to the senator: At best he's a has-been, at worst a never-was. Kerry pins great hopes on his ability to be more effective in the Senate with his new national exposure. But *Time* reported that "some Democrats on Capitol Hill privately scoff at the idea that Kerry, never particularly popular in the Senate—can expect a leadership role."[3] When the Massachusetts senator presumed to tell the Senate's Democratic minority leader, Harry Reid of Nevada, that the party had no strategy for dealing with Bush's social security proposal, Reid told Kerry that his defeat didn't leave him in a great position to give lectures on strategy. Steve Grossman, a former chairman of the DNC, put it best: "It's been a long time since the Democratic Party gave somebody a second chance. That's a big challenge to overcome."[4]

Skepticism about a new Kerry run is pervasive in Democratic circles. The *Washington Post* reported that one "well-known Democratic operative who worked with the Kerry campaign said that opposition to Bush, not excitement about Kerry, was behind the senator's fundraising success [in 2004]. 'If he thinks he's going to capitalize on that going forward, he's in for a surprise.'"[5]

The *Post* also quoted another off-the-record Democrat who was involved in formulating Kerry's campaign strategy—what little of it there was. "I can't imagine people are going to say, 'it worked pretty well last time. This is what we need next time,' he said."[6]

Democrats are, indeed, an unforgiving lot when it comes to those who blundered in national elections. But Kerry's problems run deeper than a mere lack of electoral success. He has never been very popular among Democratic primary voters. His campaign fell apart when it was seriously challenged from the left by the Internet onslaught of Howard Dean. He was rescued only by a massive attack, probably orchestrated by the Clinton people, against the Vermont governor in the mainstream media. And his one calling card—his Vietnam record—was permanently discredited by the Swift Boat vets.

Basically, John Kerry never really won the Democratic nomination in 2004; everybody else lost it. At first the nation's attention was riveted on Howard Dean, until he proved too volatile and unsteady to be president. Kerry, who received little scrutiny, beat Dean because at the last minute his status as a Vietnam veteran proved irresistible to the Democratic Party. In the next primary—New Hampshire—the media focused much more on Dean's collapse than Kerry's surge, and before Edwards had a chance to mount an effective challenge, it was too late. Lieberman and Clark, having avoided Iowa, could never even get untracked. So Kerry did not emerge from 2004 with any kind of deep base in the party. He won the nomination in a matter of weeks and then claimed the loyalty of the Democratic base simply because he was running against Bush.

If the War on Terror has wound down by 2008, Kerry is unlikely to be able to make voters believe in him a second time around.

The other 2004 leftover comes from the center of the party's ideological spectrum: North Carolina's senator John Edwards, the young, charismatic speaker and former trial lawyer who had a late surge in the 2004 primaries. Edwards was gaining support by running only positive ads as Dean, Gephardt, and Kerry slugged it out with ever more ferocious attacks on one another. For a while, it looked like Edwards could pull an upset. But ultimately, the rigged process worked just as the Democratic Party bosses intended. Though he might have beaten Kerry if he had the time, Edwards was unable to put together the resources for adequate advertising in the major primaries in March. So he dropped out early in the month and was eventually chosen as Kerry's running mate.

Going forward, Edwards does have some advantages over Kerry. One is his youth. Another is that he doesn't carry quite as much stigma from the 2004 defeat as Kerry. Indeed, in polls conducted right after Election Day, 80 percent named Hillary or Edwards as the candidate they'd most like to see in 2008 or said that they were undecided. Only one in five named Kerry, the man they'd voted for only days before.

On closer analysis, though, Edwards faces even more obstacles than Kerry in going after the 2008 nomination. For one thing, history works against him even more decisively than it does against Kerry. Only one defeated vice presidential candidate has ever won the presidency: Franklin D. Roosevelt, elected president in 1932, twelve years after an unsuccessful turn as James Cox's running mate in 1920. And only one other defeated vice presidential candidate has ever been nominated for president: Bob Dole, who ran with Gerald Ford on the Republican ticket that lost to Jimmy Carter in 1976. (Walter Mondale was defeated for vice president on Carter's ticket in 1980 and became the Democratic nominee in 1984, but he doesn't count because he actually served as vice president from 1977–1981.)

Although they have not become president, many losing vice presidential candidates have made a good impression while running on a losing ticket. Maine senator Edmund Muskie, for example, was widely hailed for his performance at Hubert Humphrey's side in 1968. For a time, he was a front-runner for the 1972 Democratic presidential nomination—until George McGovern ran him over. Joe Lieberman, who was Al Gore's VP candidate in 2000, made a bid for the presidential nomination in 2004, but lost as well. History is not in Edwards's favor.

Another bad omen is how John Edwards fared during the general election in 2004. When he accepted the nomination of his party for vice president, in a well-delivered, nationally televised speech, his trial-lawyer cadences and obvious passion resonated throughout the land. But then he disappeared. For the rest of the campaign, he was virtually invisible; any day, one expected a ransom notice from those who had taken him.

Insider gossip has it that Kerry never particularly liked Edwards. "What makes this guy think he can be president?" Kerry had asked his staff during the primary season, and his contempt appeared to carry over into the fall campaign.[7] Perhaps Kerry never felt comfortable confiding in his former rival. After all, Edwards was chosen only after Kerry had practically begged Republican John McCain to cross party lines and run as his vice president. Desperate to put a uniform on his ticket, Kerry valued McCain's genuine heroism during the long years he was incarcerated in a North Vietnamese POW camp.

After McCain repeatedly said no, Kerry turned to Edwards—more because he was the Party's consensus choice than out of any personal attraction for the North Carolinian. It was, from the start, a kind of coalition ticket between the top two finishers for the Democratic nomination—a marriage akin to the royal weddings of old, signifying an alliance between the party's left and center wings.

The VP candidate did come out of the closet briefly—but only to lose his debate to Vice President Dick Cheney. His unfamiliarity with how Washington worked (he had only been in the Senate for one term) was evident, and his inability to compete with the wise old Republican nominee was obvious. And he put his foot in it during the debate: Seeking to smear the Bush-Cheney ticket with the taint of homosexuality in order to neutralize the gay marriage issue, Edwards praised the vice president's daughter for her courage in admitting to her lesbian sexuality. It was a low blow, and the nation watched it happen on live television. They won't soon forget it. John Edwards, who attracted voters by remaining positive, had turned out to be the dirtiest campaigner of them all.

If Edwards seriously wants to pursue the Democratic nomination in 2008, he will face another serious obstacle: He's out of a job. Having given up his Senate seat to run for vice president, he has no platform from which to speak out on the issues of the day. Senators Clinton, Kerry, and Bayh and Democratic Party chairman Howard Dean all have big microphones to use in attracting the national spotlight. But Edwards has none.

Finally, Edwards's major issue—jobs—is fading with each new employment report. While we cannot rule out the possibility that a recession will grip the country before 2008—as it usually does right after a Bush is elected—an economy that is creating a quarter of a million jobs each month is not the right environment for the Edwards message. His standard stump speech about the forgotten worker who played by the rules and lost his job to foreign outsourcing—from which he never departed in 2004—may sound hollow in 2008.

The third retread who might seek the Democratic nomination in 2008 is old yeller himself, Howard Dean. Having defeated Harold Ickes, the Clinton candidate, for party chairman (a public service in itself), Dean has a splendid public platform from which to speak in the years leading up to the 2008 contest.

And that's his biggest problem. Dean has shown, from the beginning, an uncanny ability to put his foot in his mouth when he is in front of a microphone. Auditory amplification brings out the worst in him. As *Newsweek* reported, "the closer Dean got to winning the nomination, the more he seemed to misstep, to blurt out something . . . the press could hang around his neck. Dean had always been a loose cannon."[8] Dean insisted that he would not "judge" Osama bin Laden prior to a formal trial. He said that "if Bill Clinton can be the first black president, then I can be the first gay president."[9] (Presumably, he meant that if Clinton could do so much for the black community as to win their undying love, he could similarly attract gay voters.) Dean made a snorting sound during a recent speech to imitate Rush Limbaugh, whom he accused, with no evidence, of snorting cocaine. (Rush's addiction was to painkillers that he had to take following painful surgery. Like many others, he became an involuntary addict and found the habit hard to shake. He was not using recreational drugs. He got addicted because he was in pain.) And, of course, Dean capped it all by screaming like a mating chimpanzee as he conceded defeat in Iowa.

As his former guru Joe Trippi put it colorfully, "The guy [Dean] is not ready for prime time. I mean he's just fucking not ready for prime time and he never will be."[10] Dean might make a good—or, in any event, loud—party chairman, but each malapropism undermines his pretensions to the dignity a presidential candidacy requires. Very few people want a partisan attack-dog as a president. Richard Nixon had to go through an eight-year decontamination to wipe away the scent of extreme partisanship before he could be elected president. And Bob Dole never entirely lost the odor.

Another candidate, senator Joe Biden, who was knocked out of the race for the Democratic presidential nomination in 1988 by Mike Dukakis—doesn't that tell you something?—is hardly the best the Democrats can put forward. Back then, he plagiarized a speech by British Labor Party leader Neil Kinnock. Kinnock, who himself went down to defeat by one of the largest margins in United Kingdom history, was no role model, and Biden should have stuck to doing what all other politicians do—plagiarizing the work of their own speechwriters.

Biden has gained some credibility with his articulate, objective, and forthcoming analysis of the Iraq War, given from his perch atop the Senate Foreign Relations Committee. But he is, after all, another has-been who never was.

Senator Evan Bayh of Indiana, the keynoter at the 1996 Democratic National Convention, has the one qualification that seems essential for success in modern American politics: He is borrowing his last name from a more famous predecessor. Try as he might, though, he does not have the stuff of George Bush, Hillary Clinton, or even Al Gore. (Evan's father, Birch Bayh, was a legendary Senate liberal who used his power on the Judiciary Committee to bottle up all manner of Republican mischievous constitutional amendments for things like school prayer or bans on busing to achieve integration. He also authored the Twenty-Fifth Amendment to the Constitution, which outlines the procedures to be followed in the event of presidential disability. He ran for president in 1976, but lost in the primaries to Jimmy Carter.)

In other times, Bayh might be a serious contender for the presidency—and in subsequent years he still might be. But there is just no room for another centrist in the Democratic primary field. If primary voters want a liberal in 2008, they can choose Hillary. And if they want a centrist, they can choose Hillary! There is no role Bayh can play that Hillary can't play better.

Hillary Clinton's monopoly of the serious money in the Democratic Party, her husband's popularity, her control over New York State (more important than Kerry's Massachusetts, Edwards's North Carolina, or Bayh's Indiana), her lead among the super-delegates, and her rock-star appeal to women and the Democratic base all conspire to make it highly unlikely that anyone else can win the 2008 nomination.

It's Hillary's for the asking.

THE REPUBLICANS

The Republican Party is essentially a monarchic institution. Legitimacy and pedigree count for everything in deciding its nominees. While Democrats pride themselves on upsetting the traditional front-runner, particularly in the early primaries, Republicans delight in anointing their front-runner.

And here's the royal line of succession: In the beginning—1944 and 1948—there was Governor Thomas E. Dewey of New York. He bequeathed his leadership to moderate internationalist General Dwight D. Eisenhower. After a brief detour through the Goldwater right, the torch was passed to Ike's vice president, Richard M.

Nixon. Nixon named Ford. Ronald Reagan challenged Ford, and in 1980, it was his turn. Reagan begat Bush, who begat Bush (more literally)—and here we are looking at 2008.

Will the Republican Party turn aside from its tradition of following the leader and nominating someone from its party hierarchy? Not very likely.

But there is no heir. Bush has a brother, but the nation is obviously not ready for three Bushes in a row. That's too monarchic even for Republicans. Recognizing this, Florida governor Jeb Bush has emphatically ruled out seeking to be his brother's direct successor. In any case, Jeb hurt himself with the nation's political center by his doctrinaire insistence that comatose Terri Schiavo be kept alive over the wishes of her husband and despite her obvious inability to recover. Jeb may have a presidency in his future, but few think it can happen in 2008.

Cheney, the natural heir, is almost too old and surely too sick. He'll be sixty-seven in 2008—not necessarily over the hill, but getting close. And his well-publicized bouts with heart trouble make it very unlikely that he will be able to run.

No member of the Bush cabinet comes to mind (except perhaps for his secretary of state) as a possible successor.

Even if there is no heir apparent, there are front-runners in the polls—New York's former mayor Rudy Giuliani and Arizona senator John McCain. Rudy leads the field, pulling 29 percent in a June 2005 Fox News/Opinion Dynamics poll; McCain is right behind with 26 percent.[11] Both are attractive candidates who would fare well in a general election. It is even possible that these political moderates and centrists could defeat Hillary Clinton.

But neither man is likely to be nominated, despite their early leads in the polls.

Giuliani, justifiably a hero after 9/11, still has a hold on the imagination of all Americans. His forthrightness, obvious compassion, and steadfast determination as his city was devastated will long linger in our national memory.

And even before the planes crashed into the Twin Towers, Rudy Giuliani had saved New York City: When he took office as mayor in 1993, crime was eating away at the social fabric and economic base of America's largest city. With almost two thousand homicides a year, New York was no longer a favored location for corporate headquarters, the middle class was fleeing, and even the theatrical and cultural centers were feeling the pinch.

Rudy solved the city's crime problem. His gutsy and imaginative leadership of New York turned things around. Working with his deputy police commissioner, the now-deceased Jack Maple, Rudy directed the police department to approach the battle against crime as a military operation. Each day, the department's leaders would study the previous day's and week's crime statistics and redeploy their forces to cope with the emerging threats. Was there a rapist on the prowl in Queens? Did a drug gang terrorize a corner in Harlem? Did a car theft ring emerge in Brooklyn? Rudy's police department would send in reinforcements.

He insisted that the cops enforce laws against minor offenses like playing boom boxes too loud or smoking pot. It wasn't that these quality-of-life crimes were that important; his point was that they gave the police legal grounds to search suspects and find weapons and more serious drugs. The result was that criminals in New York began to find it dangerous to carry illegal weapons. They never knew when they might land in jail. He also enforced bench warrants for parole violators, in the process pulling untold repeat offenders off the streets.

President Clinton helped the city with his landmark anticrime act of 1994 (passed without any Republican help). While the bill's death penalty and gun control provisions drew the bulk of the attention, two other programs made all the difference: a provision to allocate additional funds to hire police officers throughout the nation, expanding the size of the typical urban police department by 10 to 20 percent, and an appropriation of billions of dollars for prison construction, alleviating a national shortage and giving the courts someplace to put convicted felons other than back out on the streets. In the wake of Clinton's legislation, the prison population in the United States doubled, and crime was cut in half.

But Giuliani—working with police commissioners Bill Bratton and Bernie Kerik—did even better. New York's annual homicide tally dropped from two thousand to fewer than six hundred during his tenure—and has stayed down since.[12] New York City is now rated by the FBI as the safest of the twenty-five biggest cities in the country. Its violent crime rate is now 203rd in the nation among all cities, ranking between Alexandria, Virginia, and Ann Arbor, Michigan.

With a record like that, combined with what he did on 9/11 and in its aftermath, who could resist Rudy for President? Unfortunately, a majority of Republican primary voters probably can.

Elected and reelected in the most liberal city in the country, Giuliani takes all the wrong social positions for the Republican base:

- He is a solid supporter of abortion rights and even backs late-term abortions and Medicaid funding.
- He's for affirmative action based on race and gender and energetically administered set-aside programs in New York City to encourage minority contractors.
- He strongly favors gun controls on handguns and rifles, and presided over the most restrictive gun controls in the nation as mayor of New York.
- He supports greater immigration and the delivery of public services to those who are here illegally.

In short, Giuliani flunks all the litmus tests of the Republican Right wing. At the moment, his record on 9/11 and his strong and successful fight against crime have put him ahead in the polls. A fiscal conservative, he hews to Republican economic dogma. But he is a social liberal, and the Christian Right and the National Rifle Association (NRA) control the Republican nominating process. Their standards for potential nominees are so demanding that anything less than a true believer incurs their wrath. They will eat Rudy alive in the primaries.

In 2000, for example, John McCain—who is solidly pro-life—made the Christian Right mad because he supported campaign finance reform, which, in the course of trimming the power of special interests everywhere, would reduce the power of the pro-life PACs. For this offense, they ganged up on the Arizona senator and defeated him in the pivotal South Carolina primary. These folks take no prisoners.

It is absolutely inconceivable that Republican primary voters would support a pro-choice, pro-immigration, pro-affirmative action, antigun candidate—whatever his name, outside achievements, or reputation.

As soon as Giuliani tests the waters, he will see that they are too perilous for his candidacy. He can be elected. But he can't be nominated.

Rudy should run for governor of New York in 2006. New York's current Republican governor, George Pataki, won't run again. (He says he is leaving to run for president, but the fact is that he would probably be defeated if he chose to run against New York's crusading Democratic attorney general, Eliot Spitzer.) Rudy could have the Republican nomination for the asking (New York Republicans aren't so fussy about social positions) and would have a good chance of winning, even against Spitzer.

But he's got no shot at the White House, and Rudy is too good to leave public life.

John McCain has one of the most extraordinary records in politics today. Not only is he steeped in military tradition (he is the son and grandson of Navy admirals), but his heroism is beyond dispute. Next to McCain, Kerry looks like a civilian. Imprisoned in the infamous Hanoi Hilton when his plane was shot down over North Vietnam, he suffered five and a half years of the most brutal physical and psychological torture. Never broken, tireless in bolstering his mates when their spirits sagged, he emerged from his incomprehensible ordeal tougher and stronger for it, his faith fortified and his courage proven.

No man has ever sought the presidency with a more moving tale of patriotism and fortitude.

In the Senate, he and Democrat Joe Lieberman are, in our opinion, the most honest and forthright of the members. (Is that damning with faint praise?) McCain has braved a path of independence and courage on key issues, which is rare in the upper chamber.

With the tobacco interests controlling the votes of most Republican senators with their massive campaign contributions, it fell to John McCain to cross party lines and support President Clinton's far-reaching antitobacco reforms banning advertising and promotion aimed at children. In the face of the overwhelming opposition of his own party, McCain demanded that big tobacco be subject to regulation by the Food and Drug Administration. Tobacco, he said, is an addictive drug, and cigarettes are a drug-delivery device. He failed, overwhelmed by the greed of his fellow Democratic and Republican senators for tobacco contributions to their campaigns, but his stand on principle was wonderful.

No sooner had he finished this battle than he led the successful fight for the McCain-Feingold-Meehan-Shays campaign finance reforms. While the legislation proved flawed and big money exploited its loopholes, it was the first attempt since the post-Watergate years to rein in the power of special interest money to influence American elections.

Once again McCain faced strong opposition in his own party on this score, led by Kentucky senator Mitch McConnell, the former chairman of the Senate Republican Campaign Committee. The Kentuckian wanted no changes in the old system, which had given GOP stalwarts a healthy advantage in national fund-raising.

But McCain persisted, and along with Connecticut Republican congressman Christopher Shays—the single best member of the House of Representatives—he ultimately prevailed.

(The failure of McCain's legislation to limit the power of big money in politics and the success of Dean's campaign in raising clean funds online demonstrate that the future of campaign finance reform is over the Internet, not through federal legislation.)

When corporate greed plunged America into scandal, and WorldCom, Merrill-Lynch, Global Crossing, Salomon Smith Barney, and Enron were accused of insider trading and worse, it was McCain who stepped forward with a package of real reforms. While the Senate dithered and passed a watered-down bill, McCain's proposals for real reform languished.

With a record like that, who can resist McCain? Once again, Republican primary voters can.

Universally disliked for his positions by his Republican colleagues in the Senate, McCain can count on the united opposition of his party's entire establishment. A maverick might have a chance in a Democratic primary, but in a Republican contest, he's a nonstarter.

The Arizona moderate further antagonized his party colleagues in May 2005 when he led the group of fourteen senators who negotiated a compromise in the battle over filibusters of judicial nominations. With Democrats threatening to retaliate against any rule change outlawing filibusters on judicial nominations, McCain's moderates stepped into the breach. (Slowing down the Senate's operations is like attaching a ball and chain to a turtle.) The eventual deal, which will probably come apart when Bush designates a prolifer to the Supreme Court, had each group of seven senators pledging moderation—the Democrats in refraining from filibusters except in extreme situations and the Republicans in refraining from rule changes as long as the Democrats did so.

But the deal did nothing to endear McCain to his partisan Republican colleagues who were demanding the rule change. They wanted the ability to confirm Supreme Court justices in the mode of Rehnquist, Scalia, and Thomas—extreme conservatives who could be counted on to vote to reverse *Roe v. Wade* and make abortion illegal again.

McCain's statesmanlike refusal to be railroaded by the extreme right will cost him dearly as he pursues the GOP presidential nomination.

In his failed primary campaign against Bush in 2000, John McCain demonstrated his inability to win votes in Republican primaries. Where independents were allowed to participate—in states like New Hampshire and Michigan—McCain ran well. But where they were excluded, as in New York, his candidacy ran into a wall of rejection.

For example, in the New Hampshire primary of 2000, the Arizona senator defeated George W. Bush among Independents who voted in the Republican primary by 62–19, but lost among those who had registered as Republicans by 38–41.

Since most big states preclude Independents from voting in primaries, McCain will have a very difficult time getting nominated. His contrariness and independence, the very traits that make him an attractive senator, undermine his appeal among the party faithful. Unlike Rudy Giuliani, he faces rejection not so much for his social positions—he's pro-life and antigun control—but for his general independence.

And McCain will have to compete for the Independent vote not only with his fellow Republican candidates, but with those running in the Democratic primaries as well. Most states that permit Independents to vote in any primary let them participate in whichever one they choose. So McCain would not only have to compete against Rudy Giuliani for the Independents who vote in Republican primaries, but would have to attract those who might otherwise enter Democratic primary voting booths to support—or oppose—Hillary. With her many strong supporters and equally committed opponents, Hillary would be a big draw siphoning off votes that might otherwise go to Giuliani or McCain.

In the immortal words of Yogi Berra, it would be déjà vu all over again for McCain. In 2000, the Arizona senator might well have been able to defeat Bush had not Bill Bradley, who was opposing Al Gore in the Democratic primaries that year, not drawn off Independents who might otherwise have voted for McCain in the Republican primary season.

So the two candidates whose positions are moderate and independent enough to win the general election—Giuliani and McCain—

probably can't get nominated by the stalwarts of the Republican Party.

Leading the rest of the pack of possible Republican candidates is Senate majority leader Bill Frist, a Tennessee heart surgeon trying his hand at politics. Rumored to have the support of Bush's brilliant strategist, Karl Rove, Frist has focused his campaign squarely on winning the support of the Christian Right. Rove, who knows how to win Republican primaries, probably assumes that Giuliani and McCain have the center of the ideological spectrum to themselves and needs to help Frist win the religious Right to be competitive.

Following this strategy, Frist has played chicken with Senate Democrats on the issue of using filibusters to block President Bush's more conservative judicial nominations. While the maneuver cost him dearly in credibility among his fellow senators and led to his being embarrassed by the moderates in his own party who defied him to make their own deal, it earned him plaudits from the Christian Right.

If McCain and Giuliani falter in the primaries, Frist is likely to pick up the slack.

The only problem with Frist is that he has no chance to defeat Hillary Clinton in the November election. An extreme conservative on issues like abortion, gay civil unions, gun control, the right to die, and the other hobgoblins of the Republican right, he would be Hillary's dream opponent. His candidacy would drive women in droves to the Democrats.

But Frist really does not differ much from Bush on these issues. So why will Frist alienate women voters when Bush hasn't?

When Bush ran for president in 2000, he won precisely by downplaying the issues Frist is emphasizing. Confident of winning the Republican nomination—Bush never anticipated such a strong challenge from McCain—he spoke of how he was a "compassionate conservative" committed to ending racism in the ranks of the Republican Party and restoring it as the Party of Lincoln. He said that while he was pro-life, he did not feel that *Roe v. Wade* could be overturned and felt that it should not be the key issue in the campaign. He came out against Republican proposals to deny public education to children who were here illegally. Portraying himself as a moderate, Bush was able to win in 2000.

In the 2004 campaign, Bush was much more overt about his social conservatism—but by then it didn't really matter. Terrorism was the only issue. Women who had voted for Gore in 2000 now switched to Bush because they trusted him more to keep their families safe and because they questioned Kerry's ability to handle terrorism. The war helped to reduce the gender gap among white women, despite their antagonism to Bush's social agenda.

Even if terrorism should flare again before the 2008 election, it's doubtful whether Frist could make it his issue. Rudy Giuliani—or Condoleezza Rice—would beat him hands down on the question of who would do the most to keep us safe. Without the personal security issue to bring him female votes, Frist's embrace of conservative dogma would make it almost impossible for him to defeat Hillary Clinton.

But would Frist's right-wing social agenda increase Republican turnout in November, bringing pro-life, pro-gun, antigay Christians out in greater numbers?

No. Bush already got all the votes anyone can garner from the religious Right in 2004. And those folks will be lining up for weeks before the polls open to vote against Hillary Rodham Clinton, whomever the Republicans nominate.

And don't forget: Condoleezza Rice would be just as attractive to these voters. Despite being "mildly" pro-choice, her opposition to late-term abortions and Medicaid funding and her support for parental notification and consent for abortion for minors will make her very acceptable to the pro-lifers. And her status as a self-proclaimed "Second Amendment nut" is exactly what the National Rifle Association is looking for. Finally, Condi's religious heritage and deep personal beliefs make her the ideal candidate to generate support on the religious right. But her gender, race, support of affirmative action, and "mildly" pro-choice views will assure that she won't alienate women voters as Frist would.

Even without trying to, Condi threads the needle quite nicely.

Rice would stack up well against the Republican field likely to seek the party nomination in 2008. She could compete effectively against Rudy Giuliani for the votes of those who are looking for leadership in the War on Terror. Running against Rice would put Giuliani in an odd position: His strong suit would obviously be his performance in the aftermath of 9/11, and against a Frist or a McCain, he would have the terrorism issue to himself. But Rice would nullify his strength. He couldn't hope to compete with her for leadership in the War on Terror. Her experience as secretary of state and her national leadership in responding to terror threats make her much more effective on Rudy's best issue. And Rice would have none of the drawbacks with Christian Right voters that Rudy would have.

While Condi's history obviously lacks the personal heroism of a John McCain, she would not have his baggage either. Generally well liked among top Republican elected officials and party leaders, she doesn't have McCain's reputation as a maverick. McCain's advantage would be his experience on the domestic side; Rice would have to work hard to match him issue for issue. But by drawing on her experience as an educator and an administrator, she might make McCain's legislative background appear to be less relevant to the job of president.

The list of other possible Republican candidates is as long as the fantasies and fevered dreams of senatorial and gubernatorial presidential wannabes. Nebraska senator Chuck Hagel entertains hopes of running, as does Virginia senator George Allen. Those who compose such lists often include as well New York's Pataki, Colorado governor Bill Owens, Mississippi governor and former GOP chairman Haley Barbour, and former Homeland Security secretary and Pennsylvania governor Tom Ridge. Each hopes that lightning will strike, and he will be the next president.

Despite relatively wide national recognition, though, neither Pataki nor Ridge has ever shown presidential-caliber strength in the polls. And the others face enormous obstacles when it comes to acquiring national name recognition, particularly with limited financial resources and in the face of the recognition Giuliani, McCain, Frist, and Rice now enjoy.

All of which leads us back to where we started: in the office of the secretary of state.

12

Secretary of State: What Does the Future Hold for Condi?

If Rice does decide to run for president, she would start off with two key advantages: high name recognition and a strongly favorable image. Universally known and widely liked, she would bring true celebrity to the Republican ticket and be a Republican drawing card at least as potent as Hillary is for the Democrats.

In the last analysis, however, her candidacy will hinge on how she performs as secretary of state.

None of her would-be competitors holds a job with similar opportunities for dramatic success—or risks of public failure. Rudy Giuliani is currently a private citizen. His public profile will not change much between now and 2008—as long as his erstwhile protégé Bernie Kerik behaves himself. In the Senate, John McCain has to cast difficult votes that might affect his ability to win the GOP nomination, but most senators are usually able to pick and choose their issues without getting badly hurt.

As the Senate majority leader, Bill Frist is more on the hot seat than either Giuliani or McCain. In his role as the floor leader in the Republican-controlled Senate, he must often be at the cutting edge of intense partisan warfare, which could bolster or impede his viability as a candidate.

But it is Rice who faces the most daunting challenge. Who knows how she will do as she leads America's foreign policy for the next four years? Will North Korea test a nuclear bomb—or, worse, use such a device in anger on the crowded city of Seoul, killing millions? Will Iran get atomic weapons and smuggle them to terrorists? Will Putin and Russia survive the increasing unrest in the former Soviet Union, or will a major change shake America's erstwhile adversary? Will the peace process take hold on the West Bank, as a democratically elected government begins to grope its way toward a stable relationship with Israel? And how about Iraq and Afghanistan? Will their tender democracies take root? Will the need for an ongoing American military presence, and the daily drip of casualties, ever end? Will China attack Taiwan? Will India and Pakistan come to a peaceful settlement or a shooting, and possibly nuclear, war?

And what about at home: Will terrorists strike again? Will our vigilant protectors at the FBI, the CIA, the NSA, the TSA, and the Department of Homeland Security be able to avert another 9/11?

Because we don't know the answers to these looming questions, we cannot foretell how viable a Rice candidacy might be. We can't, and Condi can't. We have to wait and see.

But the key point today is that Condoleezza Rice controls her own destiny more than any other candidate or possible nominee in either party. Others must depend on events. She can shape them. If she serves us well and the Bush second term is the diplomatic success we hope for, she will shine. But failure would dim her attractiveness.

One must be cautious. Colin Powell was highly rated at the start of his term as secretary of state, but by the time he left, his disputes with Defense Secretary Rumsfeld and Vice President Cheney seemed to dampen his popularity. His outspoken leadership in making the case at the United Nations that Iraq actually had weapons of mass destruction cost him dearly in public approval. Even if he wanted to run for president, it's unlikely that he could do so today.

Will Condi emerge similarly scathed, or will she meet this challenge in her life as effectively and skillfully as she has mastered all others?

Who knows? But, on balance, the world situation seems to hold more promise than peril as she starts on her journey.

But beware! We are incurable optimists—even as our country faces a host of daunting challenges overseas.

IRAQ

With his steadfast patience and insistence on staying in Iraq until a democratically elected government takes root and can fend for itself, President Bush has laid the basis for a significant success. While the toll of American casualties appears likely to continue, these gallant young men and women are not sacrificing their lives and limbs in vain.

Because we are in Iraq—and will not be forced out prematurely—democracy is taking root throughout the Middle East. Egypt is moving away from dictatorship. Saudi Arabia held local elections for the first time (though only men could vote), and Kuwait even extended the franchise to women. On the West Bank—the epicenter of global conflict—the corrupt dictatorship of Yasser Arafat has been replaced by a freely elected government under Mahmud Abbas.

Democracy is infectious. The purple stain the Iraqi voting officials stamped on the index fingers of the millions of Iraqis who participated in their first free balloting is the most contagious rash known to man. It is spreading throughout the world.

The very virulence with which the insurgents are battling to blow up democracy shows its potential to revolutionize not just Iraq but the entire Arab world. Rice and Bush have bet on democracy, and the wager may well pay big dividends.

Democracy is, of course, chaotic, disorganized, and disorderly. Media and politics in Baghdad will still be filled with shrill anti-American voices for decades to come. But the basic purpose of the American invasion—to replace the most vicious dictatorship in the world with a thriving democracy—seems, at last, within reach.

If the war becomes manifestly unpopular, Rice may find herself sucked down along with it. But it is up to Bush and his secretary of state to make sure that it does not become so. They both must repeatedly make the case—loudly and in public—for our involvement and connect the mission in the public mind with the dividends the war is reaping in democratic progress in other countries.

THE WEST BANK AND ISRAEL

Ultimately, of course, everyone agrees that the roots of the terrorist problem are not in Kabul, or Baghdad, or Tehran but on the West Bank. It is there that the progress must begin.

When Israel first began to build a wall to keep out Palestinian terrorists, the global community—and the United States—expressed strong reservations. There were comparisons with the Berlin Wall (even though the West Bank version was to keep terrorists out, whereas the Soviet one was designed to keep a captive population in).

But the wall has worked wonders. Suicide/homicide bombings in Israel have become a rarity. The daily, bloody attacks that once killed Israelis trying to live in peace—a kind of holocaust on the installment plan—have slowed considerably.

The wall has not only freed Jews of the dread of daily bombardment, it has made it unnecessary for Israel to retaliate with deadly force to try to kill terrorists before they can strike. The ridiculous policy of trying to deter suicide bombers by threatening to kill them has come to a close with the truce between Israel and the Abbas government concluded in the early part of this year. All truces are tenuous in this volatile region, but the wall has made it nearly impossible for Hamas and the other terrorist groups to draw Israeli blood at will.

And terror organizations need a constant flow of Palestinian blood to keep the refugee camps from which they recruit their suicide/homicide bombers seething with anger and boiling with resentment. These evil groups need a constant supply of widows and orphans to keep the embers so hot that they can flame it into violence on command.

But with a wall in their way, they have not been able to keep the region on edge, nor stoke the endless cycle of revenge killings on which terror thrives. So the attentions and energies of the people of the West Bank have increasingly gone into internal reform. They are embracing the need to clean up the notorious corruption in their PLO government, which absorbs so much of the aid the world showers on them and prevents it from reaching the needy. And they have elected a government pledged to build a viable government and nation on the West Bank.

For its part, Israel can gaze at Iraq and see more than one hundred thousand American and allied troops there, signifying our willingness to stand up for freedom and to incur casualties, if necessary, to stop terror. Taking heart from the protection the wall affords and the American presence nearby promises, Israel's politicians have summoned the will to abandon the Gaza Strip and to force the dismantling of its settlements.

A real and permanent solution to the West Bank's problems may not be far behind. Ever since the days of Henry Kissinger, American secretaries of state and presidents have longed to preside over a successful Middle East peace conference. Bringing the two sides together in a handshake that really means peace has proven an elusive goal. But the combination of Bush's decision to use force in Iraq and Israeli Prime Minister Sharon's to build a wall has created a situation that may allow peace to flourish.

This quintessential mission of American secretaries of state—Middle East peace—may well be within reach on Condoleezza Rice's watch.

IRAN

As far back as the reign of the Shah, Iran has lusted after nuclear weapons. The idea of an atomic bomb in the hands of an Islamic nation seems to fascinate the Ayatollah. Stopping Iran from getting nuclear weapons has long been a goal of American administrations, but only recently has it become an urgent priority as Iran draws closer to the nuclear club.

But there are three key points to remember:

1. Iran would not likely be engaging in negotiations with Britain, France, and Germany—and, through them, the United States—if it were planning to tell the world to go to hell and develop a bomb. (North Korea, which has done just that while in talks, had a different motivation. For the Koreans, wielding the threat of such a bomb was primarily a means to get economic assistance.)

   But the Iranians need not have agreed to talks in the first place. Perhaps they did so out of fear of what UN sanctions could do to their economy and the survival of their regime, which is manifestly unpopular at home. Or perhaps they did not really think they would ever actually manage to develop the bomb and wanted to trade the right to do so for something tangible in return. Or it is possible that the threat of U.S. troops next door in Iraq has had a sobering effect. Or they may believe that, if they go too far, an American or Israeli "surgical strike" with special operations ground and air forces might sabotage their nuclear program before it can be up and running.

   In any event, you can bet that the Iranians wouldn't be negotiating if they didn't hope either to avoid sanctions or to reap some other benefit from the talks.

2. In election after election, the Iranian clerical regime has proven to be unpopular with the people of Iran, with up to 75 percent voting for reform candidates. The ayatollahs solved the problem this year by banning any reformers from even running in the first place, but that will do nothing to endear them to their people. The Iranian regime is hardly a stable colossus that can threaten the world. The greater likelihood is that, inspired by the upsurge of democracy in neighboring Iraq—which is also Shiite—the regime will falter and fall.

   The most recent Iranian elections, of course, installed as Iran's leader Mahmoud Ahmadinejad, a man who may have been one of the hostage-takers at the American embassy in 1979. While he seems to be more of a hard-liner than his opponent, he also has about him the air of a populist who may turn Iran's preoccupation away from the "Great Satan"—the United States—and toward his nation's pressing domestic problems of low economic growth and widespread poverty.

3. Finally, even if the clerical regime does remain in place and Iran gets the bomb, it will not necessarily mean the demise of civilization. Kim Jong Il in North Korea probably does not care if his country is destroyed by a retaliatory nuclear attack, as long as he personally escapes unharmed. Saddam Hussein doubtless felt the same way. But the clerics who run Iran clearly want to build a viable Islamic state as a model for the rest of the world. They will not happily see it blown up. The conventional theory of deterrence, which held such vicious regimes as that of Stalin's Russia and Mao's China in check, will probably work here too (whether the retaliatory threat comes from the United States, Israel, or both).

The real danger with Iran getting the bomb is that they might give it to terrorists. But we already face that threat in the former Soviet Union and in the nuclear labs of Pakistan. Keeping the bomb out of the hands of terrorists is a tough job, made more difficult, but not impossible, if Iran goes nuclear.

In any case, Iran wouldn't be talking if it didn't have a reason to negotiate—and Condoleezza Rice will be a likely beneficiary of any deal that might eventuate.

NORTH KOREA

Because of his willingness to kill his own people, Kim Jong Il's North Korea poses a most significant threat. He hoodwinked Bill Clinton into signing a 1994 "Framework" Agreement that accomplished nothing more than to give him time to build a bomb in underground reactors the United States knew nothing about. By the time our intelligence learned in 1998 that Pyongyang was cheating, Clinton was mired in impeachment and in no mood for bad news.

Now North Korea has the bomb, and its leader may feel a certain impunity in using it. Since the North is within artillery range of Seoul, the capitol of South Korea and a teeming city of ten million people, Kim Jong Il can fire nuclear shells and kill millions, and we cannot stop it.

But all is not lost.

North Korea is entirely economically dependent on three countries: China, Japan, and South Korea. It cannot survive without fuel and food shipments from China or cash remittances to their starving families back home from North Koreans living in Japan. The economic ties between South and North Korea are becoming stronger every month as the regime in Seoul tries to appease the North by increasing their economic cooperation.

The question is will China use its leverage over North Korea to contain its nuclear ambitions? Ultimately, the United States has tremendous leverage to induce China to do so because we are Beijing's best customers.

The United States uses the opposite strategy from that of the nineteenth-century British Empire to maximize its global power. Britain sold its products throughout the world and made massive investments abroad. The world was its best customer, and London was its landlord.

But Sino-American relations have followed the opposite model. The United States buys $100 billion more from Beijing than we sell there, and we encourage countries like China to make large investments in American treasury bills and other dollar-denominated holdings. If Britain was the world's supplier, we are its consumer. And if London owned large parts of foreign lands, they own a sizeable portion of us.

We figure that if a country's business community makes its money from our consumers and has its money tied up in U.S. investments, they will go to great lengths to work with us. You get more leverage from buying from a nation than you do from selling to it and more from receiving its investments than from investing there. There is no more dramatic illustration of this power than our relationship with China.

The United States exports only $22 billion of goods to China while importing $125 billion from her.[1] Japan was once the trading partner that ran the largest deficit with the United States, but now its deficit is 30 percent smaller than China's.

Will the United States be able or willing to use its leverage with China to discipline North Korea? Will we be able to face the threat of the withdrawal of low-cost Chinese merchandise in our economy in order to stop North Korea from using its nuclear weapons or giving them to terrorists?

Ultimately, North Korea is really seeking an economic deal with the United States, South Korea, and Japan. They want a payoff. Their entire motivation for developing nuclear weapons was to develop a bargaining chip to trade for the food and fuel their own economy can neither generate nor afford to buy.

Eventually, the likelihood is that the willingness of America, Japan, and South Korea to satiate North Korea's economic demands and our ability to pressure China will probably lead to a solution without a nuclear attack or any needless deaths.

In the Korean Peninsula, multilateralism is the way to go. No one wants North Korea to remain a nuclear power. Certainly that is true of Japan and South Korea, its most likely targets. But it's also true of

Russia. The chance that a bomb will find its way into the hands of the Islamic militants or the Chechen terrorists who plague Russia's southern border is too great for Moscow to overlook. And China does not want another atomic power in the region to dilute its nuclear monopoly in East Asia.

With patience and diligence, the six-power talks should work out—but here even optimists must tread cautiously. The North Koreans snookered Clinton. Bush and Rice must be very careful to get a real deal before they come across with aid.

RUSSIA AND EASTERN EUROPE

When Rice first entered the White House as George H. W. Bush's expert on the Soviet Union, she watched with the world as the Evil Empire dismembered itself. Now it may happen again: The decisive votes for freedom in Georgia, Ukraine, and Kyrgyzstan may be the beginning of a vast unraveling of the former Soviet empire.

After the Soviet Union dissolved, most of its former constituent "People's Republics" became democratic in name only, usually choosing their former Communist czars as their new presidents. But with the expansion of NATO and the European Union to the edge of the former USSR's borders (and, in the case of Latvia, Lithuania, and Estonia, within them), freedom is stirring anew. Restive populations, watching the explosive economic growth in Europe and the United States, are yearning for a real shot at joining the modern world.

Former secretary of state Henry Kissinger says that Russia is either expanding or contracting; it cannot exist in stasis. With all the divergent and antagonistic nationalities subsumed within Russia, the potential for disintegration is very real and immediate. Russia is no melting pot. The various ethnic groups stick together, glaring with hostility at one another. Only Russian expansionism can keep them at bay. If Moscow is fighting on the frontier for new land, the ethnic groups it has already conquered know enough to keep quiet.

But if the spirit of revolt and democracy invades the peripheries of this polyglot empire, there is no telling where it will stop.

Vladimir Putin responded to the growing demand for democracy with more and more authoritarianism. At home, he gave himself the power to appoint the governors of the various states and eliminated the single-member districts from which half of the Duma was chosen. With all governors appointed by him and all congressmen chosen on party slates that he dominates, Putin is taking democracy away from Russia.

Abroad, in Ukraine, pro-democracy candidate Viktor Yushchenko was posioned with dioxin in an alleged effort to deny him the presidency. Despite this, he prevailed and inspired a grassroots "orange revolution." We had the honor of working with Yushchenko, and saw the passion with which the Ukrainian people have embraced freedom. This vibrant enthusiasm will not go away.

Bush and Rice could be looking at enormous opportunities to advance freedom and democracy throughout Eastern Europe and Russia.

TERRORISM

Will we experience another 9/11? Who knows? Like skilled hockey goalies (fairly rare on the ground these days, given the troubles at the NHL), the FBI, CIA, NSA, Defense Department, local police forces, and the folks at Homeland Security—particularly the Customs Service, INS, and DEA—have fielded all the shots the terrorists have attempted.

And it isn't as if the bad guys have not been trying. Dozens of attacks have been foiled. The most notable one—the attempt to blow up the Brooklyn Bridge—came to the government's attention through some of the very provisions of the Patriot Act the Left would like to repeal. But, as President Bush says, in our efforts to discover, intercept, and stop the attacks, we have to be right every time; the terrorists only have to be right once. This is an intolerable way to have to live. Our skill and luck cannot last forever.

The better course of action would be to disempower the terrorists by denying them funding. In the near term, the Bush administration has shown a willingness to adopt measures to close down charities that pass their donations through to terrorists (tactics I unsuccessfully urged on President Clinton); these have been effective. In the long term, however, the only way to avoid living under a cloud of Islamic fundamentalist terrorism is to stop buying foreign oil.

No amount of conservation will really free us of our oil habit. The current high price of fuel is due not to any disruption in supplies, but to the ongoing, increasing demand for oil in much of the previously unindustrialized third world—primarily in China and India. The oil market will only get tighter. Opening up Alaska will be a drop in the bucket as long as we import more than twelve million barrels every day; 60 percent of our petroleum comes from abroad.[2]

The real answer to terrorism is to switch our cars from petroleum to hydrogen. These hybrid cars actually run on electricity, but their battery is recharged while the car is in motion by a fuel cell running on hydrogen.

Hydrogen leaves no trail of pollution and causes no global warming. And using it does not fund the terrorists who would destroy us. Hydrogen is not a green issue—it's red, white, and blue. Bush and Rice could fatally weaken the terrorists in one stroke by moving ahead with a hydrogen program for America, helping to ameliorate global warming in the process.

We could generate hydrogen either through electrolysis of water or by a noncombustion chemical process using natural gas.

We have lots of water. Even sewage could be used. Electrolysis, of course, would use up a lot of electricity, but the vast bulk of our power comes from coal, hydro, and nuclear plants. Oil accounts for very little. (Two-thirds of our national demand for oil goes to feed our cars.) So an increased need for electricity need not make us dependent on foreign oil.

Of course, we would rather not have to burn more coal, but economies of scale and better technology should reduce our need for electricity to make hydrogen from water in due course. And coal burning need not lead to more pollution or global warming. As Terry Tamminen, prominent environmental activist and former secretary of environmental conservation in California, points out, carbon dioxide produced by coal burning "can be sequestered back into the coal seam itself" through current technology without escaping into the atmosphere.[3]

Or we could get our hydrogen from natural gas. This would involve relatively small amounts of pollution, particularly compared to oil or coal. We can get natural gas (or methane as it is called) from "farm waste, urban green waste, and recovered landfill gas," according to Tamminen.[4] Until we generate enough from these sources, we can import natural gas in liquefied form from plenty of suppliers—like Australia, Chile, and the rest of South America—who have nothing to do with OPEC or terrorism.

But the Bush administration has a blind spot—as one would expect of a Texas oilman—when it comes to moving away from petroleum. While the president has embraced hydrogen as a long-term solution, we need immediate action.

And California governor Arnold Schwarzenegger is leading the way. The Governator has set in motion a state program, supplemented by federal and private funds, to produce enough hydrogen and offer it at enough service stations to permit Californians to switch en masse to cars powered by hydrogen fuel cells by 2010.

Schwarzenegger plans to require gas stations along California's interstate highways to offer hydrogen and to produce enough of the gas to power California's cars. Since almost everyone there lives near one of these roads, it will be possible to drive a hydrogen-powered car throughout the state. He plans to team up with British Columbia, Washington State, Oregon, and Baja California to create a "hydrogen highway" that would run from BC (British Columbia) to BC (Baja California).[5] Since one-fifth of the new car sales in the United States are in California, the tail may soon wag the dog into doing what we need to fight terrorism.

Of course, hydrogen is also the answer to global warming. Just how long does the Bush administration plan to keep its head in the sand, deny that global warming is happening, and postpone taking any serious measures to deal with it? How much does the Arctic ice cap have to melt or Antarctica dissolve before his administration takes it seriously?

Bush doesn't seem to understand how serious a threat climate change—and the rising sea levels it causes—really is. Virtually the entire planet is united on this issue—so much so that the Kyoto Treaty is likely to take effect without American participation. But the United States is almost alone in its refusal to recognize the obvious—that something is drastically, dramatically, and dangerously wrong with our global climate.

Conservatives in the United States agree that global temperatures are rising, but say that it is due to natural causes like volcanic eruptions about which we can do nothing. Moderates and liberals agree that volcanoes cause the bulk of the warming trend (by sending up soot and ash that traps the earth's warmth like a blanket), but argue that we need to reduce man-made emissions to compensate for it anyway. We cannot simply surrender the homes of hundreds of millions of people to inundation. The tsunami that afflicted South Asia will seem mild by comparison to the hundreds of millions who will be flooded by rising sea levels over the next fifty years.

Bush and Rice do not need to reverse themselves on the Kyoto Treaty. They just need to follow the Governator's lead.

So Condoleezza Rice faces great challenges as she assumes the daunting responsibilities and thankless tasks of secretary of state. But there seem to be clear answers to many of our most pressing problems. Indeed, the effectiveness of the Bush military action during the first term has opened the way for tremendous progress on many of the most important fronts of American foreign policy.

As an integral part of the team that developed Bush's first-term foreign policy and made these openings possible, Rice is obviously alive to each of these opportunities. We cannot tell how well things will work out, but it is at least a strong possibility that her reign at State will be a happy one, filled with important progress for our country.

If Rice has a strong hold on public affection now, it is likely to get better and better as we watch her at work in her high-profile job. It's certainly safe to assume that none of her opponents, least of all Hillary, is in a position to improve his or her standing, popularity, or profile as dramatically as Condi.

But if her supporters want Rice to run, they can't just sit around waiting for it to happen. By definition, a "draft" is a spontaneous outpouring of support, generated not by the candidate but by her backers, eventually becoming so intense that she has to yield to its enthusiasm and run. Her fans must do a great deal of work to make it possible for Condoleezza Rice to run for president. Candidates like Condi cannot be drafted without a real effort at the grassroots level.

Here's what her backers will have to do. . . .

13

Drafting Condi

If Condoleezza Rice is to run, a candidacy must be built around her, by her supporters. It must come together without her efforts, put in place by those who want her to be president.

So how would a draft work? With no smoke-filled rooms to beckon a dark horse to enter the fray, can modern voters create a candidacy and then invite the candidate to join them?

Not only can it be done: with the Internet and the modern communications tools now available, it will be easier than ever to accomplish.

Her supporters must think of her candidacy as a bagel: filling in the surrounding campaign and leaving the hole vacant until the candidate shows up. They must construct a campaign with sufficient funds, support, geographic spread, and momentum before there is a candidate to catalyze it. Like nutrients that surround the embryo, the campaign needs to be a thriving affair before the candidate can possibly run.

How would the momentum develop for such a candidacy? It would obviously begin with the message that Rice is the strongest alternative to Hillary. As the former first lady nears the finish line and comes close to winning the Democratic nomination, those who know what a terrible president she would be will increasingly see Rice as the obvious antidote.

At the same time, as the candidacies of Giuliani and McCain fade, dampened by the rejection of the GOP voter base, it will become obvious that the party is at risk of nominating someone who cannot possibly defeat Hillary—another white, male, pro-life, right-winger, exactly the kind of opponent Mrs. Clinton would love to run against.

The demand for Rice will rise from the grassroots (and cyber roots) as more people come to see the obvious need for her to run. At the same time, Rice's accomplishments as secretary of state and her performance on the most demanding of global stages will animate her image and enhance the prospect of her candidacy.

As the national media focuses on the idea of a Condi candidacy, "Condi Clubs" will spring up throughout the nation. Without a candidate, or even a high-profile surrogate, the movements will initially appear chaotic and uncoordinated—but that's what happens in genuine grassroots campaigns.

The "Rice for President" campaign need not be centrally controlled. It need only have coherence at the very local level. Her supporters must not wait for orders, but need to take the initiative, thinking nationally but acting locally.

Even though a presidential campaign seems to be a national affair, it doesn't start out that way. Long before Election Day, years before the nominating conventions, presidential campaigns are really run on the state level. Since each state has its own primary or caucus system run by its own rules, a presidential campaign initially forms on the local level. States that select their convention delegates early in the process—like Iowa, New Hampshire, Delaware, Arizona, and South Carolina—assume paramount importance. But right on their heels are the large states, now frontloaded in the nominating process of both political parties—like California, New York, Florida, and Texas.

Normally, a presidential candidate would take the lead herself in generating local committees in key primary states, touring them, hosting fund-raisers there, and contacting key party activists to recruit them before another candidate comes calling.

But with a draft movement, her supporters must assume that Rice will never pick up the phone to ask for backing; her supporters have to do it for her. The local organizations that are essential to any presidential race will form spontaneously, taking on a life of their own.

But even a state organization is too big a place to start. Most states choose their convention delegates at the congressional district level. Each district sends its own delegation to the state and national conventions, and its voters decide for themselves whom their delegates will support—at least on the first ballot. So Condi's supporters will need to organize in each of the nation's 435 congressional districts—and in Washington, D.C.; Guam; and the other areas that are not represented in Congress but send delegates to the convention.

In many states and districts, the delegates supporting Rice may have to run as "uncommitted" slates, since some jurisdictions require that a candidate affirmatively take steps to be listed on the ballot—something Rice won't do if she hasn't decided to run. But the very drama and newsworthiness of "uncommitted" slates winning primaries and drawing support in polls will attract coverage and make the Rice candidacy unique and enticing.

While other candidates desperately fan the fires of passion for their race, Rice will do nothing—but her candidacy will grow daily as people flock to her banner.

As the Rice for President Committees—the Condi Clubs—nominate delegate slates in each congressional district, they will have to raise money for their campaign. Again, they have to act locally. Even though announced candidates are pouring PAC and special-interest money into the race, the Condi Clubs will have to generate their funds from neighbors in their region and online supporters elsewhere. Each committee, or club, needs, and will likely get, enough money for its own campaign, much as the Dean operation did in 2004.

When people phoned or e-mailed the national Dean headquarters, they asked what they could do to help. The staff person at the other end of the phone would just say: "Go out and talk to your network of friends, family, and neighbors about Dean." How they did it was their job to figure out. In the Rice campaign, there may not even be anyone to ask.

Rice won't need as much money as her competitors in the Republican primaries. As secretary of state, she will already be the object of massive media coverage on a continuing basis—racking up more free publicity in a matter of weeks than many candidates could muster in the course of an entire campaign.

And this much seems certain: As Rice jets around the world, she will take note of the groundswell of support for her candidacy—if only from the corner of her eye.

If Rice were invisible in her day job, it would be hard to sustain the momentum for her candidacy. But her visibility will likely outstrip the most industrious of candidates as she carries out her duties at the head of the State Department. Her every accomplishment and success, every speech or pronouncement, will further fuel the fires of her supporters back home.

Rice will have to seek to squelch the fires. She not only must avoid being seen to encourage her backers, but also must be visible in restraining them and dousing their enthusiasm. She may even mean it when she says that she doesn't want to run. But the decision is not hers to make. A democracy can choose its own leaders.

The question is: If nominated, will she run? If elected, will she serve?

Rice's opponents will find her an elusive target as they swing blindly against a phantom candidate who does not appear in debates, declare her candidacy, or run television ads. As they vie for support, they will come to find meaning in the famous children's verse by Hughes Mearns:

As I was going up the stair
I met a man who wasn't there.
He wasn't there again today.
I wish, I wish he'd stay away.[1]

They will find themselves running against not a candidate, but a group of voters who are boosting a possible nominee. Politicians have little difficulty in attacking a candidate, but they find it very injurious to attack the voters who back her.

Rice, like Eisenhower in 1952, can float above the field without dirtying her hands or either taking or landing punches.

In some states, candidates must file their intention to run in order to get on the ballot. But in others their names are automatically included unless they decline. In still others—such as California—the name of any generally recognized candidate is automatically included on the ballot.

The first political event of the year—the Iowa caucuses—poses little problem for the draft Condi movement. Those who attend these meetings, held throughout the state, can vote for whomever they want, with or without the candidate's permission.

But in New Hampshire, Delaware, and Arizona, a candidate must affirmatively file a "declaration of intent" before her name can be listed on the ballot.[2] Since Rice probably won't file one, her supporters will have to run as uncommitted delegates and make clear that they are really pledged to support Condi. In Florida and New York, voters can submit names to the secretary of state, who then forwards them to the Presidential Candidate Selection Committee. For a candidate to withdraw, she must file a written affidavit saying that she has no intent to become a candidate.

Even if Rice files such a statement, however, delegates can still run as "uncommitted"—and get elected anyway.

If one of these uncommitted slates wins an early primary, signaling a boom for Condoleezza Rice, there will be no stopping her candidacy!

Will enough money be available? The massive grassroots enthusiasm for her practically guarantees it. And any lack can be made up by the vitriol of those who want to stop Hillary Clinton from becoming president. Together, these two forces will tap into more than enough money to propel a Rice candidacy . . . even without Rice.

How will the mainstream media treat the draft Rice movement? Initially, it will assume that unless a campaign starts with a candidate, it cannot succeed. Having never seen a draft, they will declare it unworkable, pure fantasy.

But the fires of the Condi movement will not be so easily doused. As local activists gather together to form clubs, recruit supporters, and nominate delegates, the enthusiasm for Rice in each individual community will become more manifest. Card tables will spring up in front of local shopping centers, movie houses, laundromats, athletic stadiums, and the like every weekend. Volunteers will pass out homemade literature and write down e-names to augment their lists. The snowball will grow as it descends the hill.

As happened with the Dean candidacy in 2004, the media will "discover" the draft Condi movement, and its very novelty will attract increasing coverage. Finally, when the absent Rice begins to rise in national public opinion polls, the phenomenon of the candidate who wasn't there will come to define the Republican contest.

Meanwhile, the alternative media, so pivotal in the Bush campaign of 2004, will go to work promoting Rice. Her candidacy will become the subject of talk radio, call-in shows, NewsMax releases, and coverage on Fox News and the other cable stations. The same channels that propelled the accusations of the Swift Boat veterans into national prominence will make Rice's candidacy a topic of everyday street corner discussion. Fanned by her aggressive and successful efforts to do America's work abroad, Rice will gain momentum without ever lifting a political finger.

We have a precedent for drafting a presidential candidate: the campaign for General Dwight D. Eisenhower in 1952. In that year, New York governor Thomas E. Dewey and Massachusetts senator

Henry Cabot Lodge successfully mounted a campaign to draft Eisenhower to run for president as a Republican.

Oddly, four years before, in 1948, Ike had been approached by the *Democratic* Party to run as its candidate in place of President Harry Truman, who most party leaders doubted could win reelection. Though they were unsure of Eisenhower's political leanings, the Democrats realized that he would make a good replacement for the beleaguered president. Eisenhower turned down the overtures in 1948, but they surfaced again in 1952—this time from the Republicans.

But no one knew if Ike was a Republican and if he was willing to run as the party's candidate for president. At the time, General Eisenhower was serving as the first commander of NATO. He maintained a strict silence on domestic American politics, saying that it would be inappropriate to speak out while he was still in uniform.

Would Eisenhower run? Herb Brownell, his future attorney general, writes, "Ike never put himself in the position of saying he would not run." He provided "absolutely no encouragement" to those trying to draft him into the race, but "he never said no absolutely" and never took the Sherman oath.[3]

It was not until January 1952, on the eve of the New Hampshire primary—then, as now, the first in the nation—that Eisenhower even revealed that he was a Republican and would accept the nomination if it were offered. But even then Ike said that only "a clear-cut call to political duty" would induce him to resign his post at NATO, which he called "the vital task to which I am assigned."[4]

As with Eisenhower, it is the task of Condi's supporters to convince her that they want her to run for president. No matter how much they are motivated by belief in her and how much by the desire to prevent President Hillary, the American people must reach out to Rice to ask her to run.

Eisenhower did not have a free ride to the nomination. The Republican floor leader in the Senate, Robert Taft of Ohio, was the overwhelming favorite to win his party's presidential nomination.

Known as "Mr. Conservative," Taft was an isolationist who opposed any involvement in Europe and most engagement with the rest of the world.[5] In 1940, he had declared that a German victory was preferable to American involvement in World War II. He was best described by *Forbes* as "one of that vast group of Americans to whom other countries seem merely odd places, full of uncertain plumbing, funny-colored money, and people talking languages one can't understand."[6]

But Taft had won the heart and soul—to the extent that either existed back in those days—of the Republican Party. In polls, Eisenhower, still not a candidate, was favored by the Independent voters the party had to attract to win, but Taft was still the *beau ideal* of the GOP stalwarts.[7]

And yet the Republican governor of New York, Thomas E. Dewey, who had lost to FDR in 1944 and Truman in 1948, kept pushing for Eisenhower. Ike, he felt, was "the only Republican who could regain the White House in 1952 'if we could get him to run.'"[8] But, as with Condi, it was by no means certain that Eisenhower would enter elective politics. As future attorney general Herb Brownell recalled, "I had observed that Ike had never definitively closed the door on accepting a Republican nomination and at the same time had preserved his image as a noncandidate."[9]

Like Rice, Eisenhower was extremely well informed about foreign and global affairs, having led the Allied forces in Europe in World War II. But his record on domestic issues was less clear. Eisenhower endorsed the basic Republican principles of a balanced budget and a free enterprise system as devoid of government regulation as possible, but only elaborated his domestic policy opinions after winning the nomination of his party.

And, then as now, it was international policy that commanded the nation's attention. Locked in a drawn-out war in Korea with no end in sight, challenged at every turn by a Soviet Union still in the grip of Stalin, Americans were as focused on foreign policy as they are today.

Many shared Dewey's view that Eisenhower could win—and that only he could—but wondered if a campaign could be waged without a candidate. Eisenhower, for his part, was determined not to get mixed up in the unseemly business of seeking the nomination unless he had evidence of a broad public demand for his candidacy.

The key to Eisenhower's strength lay in grassroots support, organized into "Citizens for Eisenhower" clubs. Brownell reports that "The Citizens group was made up mostly of volunteers who had had no experience in organized politics. They believed that volunteers, by expressing their opinions openly, could force the national convention to act."[10] Even without a candidate, the draft-Ike movement entered the general's name in the New Hampshire primary. The state's governor, Sherman Adams (later Eisenhower's chief of staff), backed the noncandidate, and the Eisenhower forces descended on New Hampshire for a pitched battle. Taft battled hard to win, but could not beat the popular general. Eisenhower won by 46,661 votes to 35,838.[11]

How did Eisenhower win without even campaigning? He realized that staying at his NATO post, defending Europe against Communism, was a far more effective advertisement for his competence than shaking hands in snowy New Hampshire. As Ike wrote Dewey, "I keenly realize that I am of no particular help in all the matters of policy and decisions with which you people [the draft-Ike campaign committee] are continually faced. But I feel . . . that as long as I am performing a military duty and doing it with all my might, I am possibly providing as much ammunition for your guns as I could in any other way."[12]

Just as Rice would be providing her backers with ammo by being a good secretary of state.

The Eisenhower campaign, sans Ike himself, gained momentum. In Minnesota's primary, it had to face the state's former governor and hometown favorite, Harold E. Stassen. Because local rules required that only declared candidates could be listed on the ballot, the Eisenhower forces had to wage a write-in campaign—a daunting task. But Eisenhower's popularity overcame even this obstacle, and the general racked up 108,000 votes, almost as many as Stassen's 129,000 and far more than Taft's 24,000.[13] The Minnesota result was astonishing. To go into a rival candidate's backyard, not appear on the ballot, do no campaigning, and get almost as many votes as the former governor of the state was a political tour de force. From then on, when people asked about Ike's support, Dewey would answer "certainly it was not gremlins who wrote in his [Eisenhower's] name on all those ballots in Minnesota."[14]

Money poured in, even without the candidate chipping in to beg for it by phone or pump hands at receptions. A smoothly running public relations machine blanketed the nation with pro-Eisenhower material, and Ike's day-to-day work as the first NATO commander in shaping the West's response to the Russian threat gave the general all the publicity he needed.

Eventually, just one month before the convention, Eisenhower resigned his military commission and announced that he was running for president.

Then Taft made a mistake. In solidly Democratic Texas, the regular Republican organization was so small it could meet in a phone booth. But the Eisenhower candidacy was bringing many Independents and former Democrats out of the woodwork. When 75,000 of them invaded the Republican caucuses held throughout the state to rubber-stamp the party leaders' choice, Robert Taft—and voted instead for Eisenhower—the bosses panicked. Even though these Eisenhower enthusiasts complied with state law and signed "loyalty oaths" promising to vote Republican in November so that they could vote in the caucuses, the party leaders refused to count their votes.[15] Anyone who did not vote Republican in 1948, they said, could not participate. The Eisenhower forces cried foul and bolted the caucuses, sending a rival delegation to the Republican National Convention in Chicago. Taft forces used similar tactics to stem the flood to Ike in other Southern states, where the Republican Party was only an empty shell without the Eisenhower recruits.

The national media and public opinion were outraged at the high-handed tactics of the Taft forces, and the convention delegates themselves voted to overrule the state party hierarchies and seat the Eisenhower delegates.

Even so, the battle went down to the wire at the Republican convention. At the end of the first ballot, Eisenhower led with 595 votes; Taft had 500, and the other candidates divided up the remaining delegates.[16] But Ike was nine votes short of the tally he needed to carry the nomination . . . until his erstwhile rival, Harold Stassen, stood up and switched enough votes to Ike to give him the nomination on the first ballot.

If Eisenhower could do it in 1952, why can't Condi in 2008?

And what will President Bush do while all this is going on? He'll look on with amusement, surprise, and, one suspects, increasing satisfaction. After all, Condoleezza Rice is most purely his creation. Of all the candidates in the race, only she comes from his circle. Giuliani was mayor of New York and a hero of 9/11 long before he ever embraced Bush. McCain first came to the national political scene as Bush's opponent in 2000. Frist is cooperative with the president's agenda and unquestionably his personal choice for majority leader in the Senate (after Bush pushed Trent Lott aside), but the president is too good a politician not to see the limits of his candidacy.

But Rice is in a totally different category. She is a personal friend of the president, his representative to the rest of the world. Her power as secretary of state comes from her unquestioned access not just to Bush's office but to his mind. Long before he was president, as they worked out on adjacent stationary bikes together and talked football and foreign policy, these two people have formed a bond that is strong and lasting. She has practically become a member of the family, almost an adopted little sister to George and Laura.

If President Bush is convinced that the draft Condi movement is real and realizes that she is doing nothing to fan it, he may become an increasing enthusiast. As with Condi, one cannot expect him to create the campaign—but if you build it, he may come, too.

At what point will Rice actually leave her post and join the fray as a declared candidate? When does a guerilla army stop using hit-and-run tactics and stand up to fight a conventional war? According to precedent, from George Washington to Chairman Mao, it does so when it acquires sufficient strength among the people to be able to go head-to-head against armies of great conventional strength with a hope of victory.

For Rice, the moment will probably come—if it does at all—after some surprising initial success in early caucuses and primaries. One can see her remaining at her post throughout all of 2006 and 2007, until the Iowa caucuses and the New Hampshire primary in January and February 2008. If she scores well in those contests, especially if she wins one or both of them, the popular momentum will be unstoppable. At that point, one can easily imagine Condi resigning as secretary and jumping into the race with both feet.

But at that point, it will be too late to qualify to run as a Rice delegate in most states. Rice's delegates will have had to make it onto the ballot on their own, often officially filing as uncommitted delegations. Then, when Condi begins to win the early primaries and caucuses, they will already be on the ballot.

So the draft Condi movement needs to have active supporters in all fifty states, and file complete sets of delegates—committed to Rice or, where legally necessary, officially uncommitted but in vocal support of Condi—so that Rice will have a platform of support if the draft movement's initial successes encourage her to run.

It is more important that the draft movement have donors than donations. The actual need for cash will be limited in the early days of the 2008 campaign. With no candidate to tour, there will be no need to pay for travel—one of the biggest early expenses in any campaign. Printed flyers and television advertising will be concentrated in the few early states—perhaps only in Iowa and New Hampshire, where media is inexpensive and easy to buy. The real money won't be necessary until February 2008, after Iowa and New Hampshire have voted, when the campaign must gear up for the big state primaries in March in New York, Florida, Texas, and California.

By then, Rice will likely be in the race. While it will be too late at that point to begin to prospect for donors, it will not be too late to send out an emergency e-mail to those who have already signed up to raise the $10 million or so needed for the March primaries. The key is to have a list online and in hand from which to solicit.

The turnaround time for online donations is instant, and there is no big upfront cost for postage or manpower. So the campaign can reload quickly and frequently as it moves from its guerilla stage to the period of conventional campaign warfare.

After that, Condi can take care of herself. Her vast experience in front of cameras and microphones will work to her advantage once she emerges as a regular candidate facing her rivals. But the glitter of being the first candidate to be drafted in fifty years will linger. Hers will always be the people's campaign, created by popular demand, not by personal ambition.

It will be a phenomenon without precedent in our modern world, one that will fit precisely the new spirit of activism and involvement kindled in the wake of 9/11. For many, turned off by the domination of the political process by special interest and professional politicians, it will represent a triumph of the average man and woman and a new demonstration of his and her relevance to our political system.

There is no better way to launch a candidacy than in response to pressure from the grassroots. That's why every candidate pretends to hear those voices as she steps up to the microphone to announce her quest. But the voices are usually only faint and distant. And more often than not, they are artificially created and amplified. There are no mobs in the streets demanding that she run, only her own ambition.

But with Condi it will have been different. And that gap—between the appearance of a grassroots demand for a candidacy and the obvious reality of a genuine movement—will fuel her entire endeavor.

14

President Clinton? President Rice? What Kind of President Would Each Make?

So what kind of president would Hillary be? How would Condi handle the job?

The presidency is a very, very complex job. Look at a candidate's public positions, and you can pretty well figure out what kind of governor or senator they'd be. The presidency, however, is a job where temperament, insight, charm, diplomacy, and crisis management skills—along with policy decisions—determine success or failure.

Let's start with policy. Hillary Clinton would be the most liberal president we have had since Lyndon Johnson. Bill Clinton is a moderate by choice and, sometimes, a liberal by necessity. But his wife is the exact opposite. She tries to be as liberal as possible as often as possible and embraces centrism only when political reality makes it necessary. Hillary believes that government delivers services well, and that the quest for private profit is the root of all selfishness and vice in American life. She seethes in anger when insurance brokers add to the price of health care to make their commissions. She boils over when the rich want to pay less in taxes and won't step forward to fund a full range of services for the poor.

If the collapse of socialism and the success of Reaganism have educated many former liberals (including us) to the defects of the public sector and the advantages of harnessing the profit motive to help the poor, Hillary has learned nothing. She is a dedicated proponent of expanded public services, particularly where women and children are concerned.

Hillary came of age as a student leftist and grew to maturity as first lady of Arkansas. She still echoes the concerns of Little Rock liberals that taxes are too low in their state and that the investment in public infrastructure—schools, hospitals, roads, and such—is too attenuated. Hillary would raise taxes to generate revenue, but she would also raise them because she believes in a larger public sector. She sees taxation not just as a means to fund government, but as a way to mitigate the disparities of wealth and achieve a more egalitarian society. She wants taxes not just to pay for programs, but to redistribute income downward. Hillary believes deeply in income redistribution; she sees the federal government as a cash transfer machine, taxing the rich to aid the poor.

Bill Clinton is in politics to be liked. His neurotic narcissism demands that he prove his self-worth to a doubting superego by finding more and more people who like him, each a mirror in which he can see a good version of himself. His desperate need to win elections stems not only from his love of the perks and privileges of power—and his genuine desire to use it to help people—but from the psychological balm that public approval rubs on his poorly-formed self-image.

Hillary, on the other hand, does not care if people like her. She is not a narcissist. She believes deeply that she is a good and moral person, and she is so convinced of her own virtue that she does not particularly value the opinions of others. She is a woman on a mission—a public policy mission—to bring liberalism back to the mainstream of American life and to expand the role of government in helping the poor and middle class.

On occasions when Bill's policies proved unpopular, he was willing to change them. Hillary never will. She will persevere right to the end, as she did with health care reform, and go down fighting if she must. In this respect, she has much more in common with the current President Bush than she does with his predecessor. Her inner sense of her own rectitude has bred in her a stubbornness that dominates her political outlook.

But for all her rigid liberalism, once in power she is capable of endless subterfuge in her efforts to attain higher office. She will run as a moderate in order to govern as a liberal, controlling her steadfast liberal impulses until she has secured the office from which to indulge them.

In foreign affairs, Hillary's views are less clear. She is only just learning about these issues, and her real opinions have yet to emerge—and probably won't until after she is elected. During her White House years, she was a peacenik, opposed to foreign interventions, against American involvement in Somalia, and concerned that the administration would be too preoccupied with the Balkans. But now, who knows?

For her part, Condoleezza Rice shares the basic Bush/Republican outlook on public policy issues. She would likely seek to hold down taxes, limit the role of government, and harness the private sector for the delivery of public services. She probably shares with Margaret Thatcher the perception that "private vice makes public virtue"—that the desire for profit and a better life makes the economy work for everyone.

Like Hillary, Condoleezza Rice is a woman on a mission. But Rice's mission is the expansion of democracy. Where Hillary would focus primarily on expanding the role of the government at home, Rice would want to see us become more involved abroad. A dedicated adherent of Bush's fixation with spreading freedom throughout the world, Rice would bring a Wilsonian vision to the Oval Office, working to chart the way toward a world of democracies.

But one senses that the stubbornness of her ends would not bleed over into inflexibility in the choice of means by which to achieve them. Rice is first and foremost a diplomat, skilled in handling people and knowledgeable about the byplay of conflicting interests and goals. Her long experience in managing foreign affairs would likely lead her to seek out common ground and to advance one step at a time. She would probably bring a Rooseveltian flexibility to how she advances her ideas, coupled with a Bush-like tenacity in focusing on the ultimate goal.

But it is in temperament that these two women would most differ in the presidency.

Rice hates to make enemies. While she is quite capable of standing her ground in any debate or give-and-take, she tries to charm rather than compel, to finesse when others would confront. She seems to get angry rarely and appears to keep a balanced perspective even in the heat of combat. While she identifies her enemies, they tend to be America's adversaries, not her own political opponents. Her past is not littered with examples of getting even or bearing grudges. She doesn't seem to have an enemies list, but gives every appearance of internalizing her Christian faith in her dealings with her colleagues and competitors.

As first lady, Hillary Clinton defined herself by her enemies. She needed her adversaries constantly in her sights to reassure her that she was doing the right thing by opposing them. As long as she is making the wealthy and privileged angry, she figures she's on the right track. Like a dolphin that bounces sonar off the ocean bottom to determine its position in the sea, she uses conflict and adversarial confrontation to make sure she is pointing in the right direction.

Without conflict and enemies, Hillary was subject to a debilitating entropy and confusion during her years as first lady. She wallowed without direction in her efforts to design a good health care system for America, for example, but sprang forward with zeal and alacrity once the issue was framed as an effort to cut back the funding of health insurance providers, brokers, and wealthy doctors whom she felt were overcharging for their services.

And now? Has Hillary's Senate tenure led her to use the politics of alliance rather than that of conflict to build support and advance her agenda? It's hard to tell. We can only hope that the growth she shows in replacing anger with charm and criticism with a search for common ground is real.

One sees something of Eisenhower's indirection and Reagan's charm in Rice's style. But there are grounds for concern that Hillary has too much of Richard Nixon about her. In the White House, Mrs. Clinton approached life and politics with a massive chip on her shoulder, which grew larger amid all the political and personal attacks she had to face. Back then, one could well imagine her sharing FDR's view: "The economic royalists hate me and I welcome their hatred."[1]

The Hillary of the White House years allegedly used the FBI, the IRS, and all the implements of government to get even with her enemies. She gloried in the process and did it with belief in her own virtue. And where public agencies are too limited in their ability to harass her adversaries, she might resort, as she did during her White House years, to her own ad hoc force of private detectives and lawyers to do her dirty work. She could reinvent what we have called the secret police to destroy those who stand in her way.

But how much of Hillary's paranoia stemmed from the fact that it was Bill for whom she was covering? What part of her bitterness stemmed from her role as understudy and supporter rather than as principal? Would she grow, as Bobby Kennedy did, into a better person as she steps out of the shared spotlight onto the stage as her own person?

If not, Hillary would quickly find herself isolated in Washington, without the support she would need to govern. Members of the media, held at arm's length and manipulated by sleight of hand when they are allowed access, would come to dislike her intensely, and their animosity would quickly spread throughout the capital. Members of Congress, even those who would normally be her allies, would find themselves enraged by the petty slights she is capable of and cease cooperating with her.

One senses that Condoleezza Rice likes staff members who offer constructive criticism. Reared in academia, she seems to relish advice and seek out counsel. Like Bill Clinton, she wants to know what other people think of her and constantly seeks feedback.

But Hillary does not. Many times I have sat with her top staffers, trying to figure out who would have the guts to go into the lion's den and tell Hillary she was making a mistake. Often I assumed the role myself, only to emerge with my head in my hands. The internal checks and balances that can pull a president back from extreme positions and danger are almost entirely lacking in Hillary's makeup.

Rice doesn't lie. She has been on the public stage for so long that she must realize how easily one can get caught, and as a result, she probably prefers the traditional Washington art of circumlocution where others might just make up a story.

Until a few years ago, Hillary appeared to find it almost impossible to avoid making up yarns to suit the convenience of the moment. Today Hillary's staff works overtime to script her and keep her under control, precisely because they know how easily she can be tempted into fabrications big and small. And their controls seem to be working: The off-the-cuff exaggerations and fibs seem to be diminishing, and Hillary appears to have a new discipline about her. Will this process grow as she approaches the goldfish bowl atmosphere of the American presidency?

Rice has never been materialistic. While always stylishly dressed, she was enough the product of a bourgeois upbringing to resist the allure of acquisitiveness. In her superintendence of Stanford's $1.5 billion budget and her use of the perks available in the White House and at the State Department, she has managed to avoid any hint of ethically questionable conduct and appears to live easily within her salary and savings.

But Hillary has a sense of entitlement that leaves her vulnerable to the temptations of financial misconduct. So convinced is she that she has led a life of sacrifices as a dedicated public servant, that her record is filled with shortcuts and the cutting of ethical corners. Her personal appropriation of gifts intended for the White House, her use of Bill's power to get a retainer from Jim McDougal's Arkansas bank, her acceptance of highly questionable advice and favors from those seeking favors from her husband to steer her to big winnings in the futures market, all attest to her willingness to take risks to make extra money.

But now she is rich. Her $8 million book deal and Bill's $12 million deal, along with his massive speaking fees, have changed the financial calculus of deprivation that may have motivated Hillary's past financial chicanery. Today she is much more than financially secure—she is wealthy. That change in her financial status may eliminate any recurrence of her previous ethical problems.

And how would each woman respond to a crisis as chief executive? Both would probably handle it well. Hillary, who has the discipline that Bill so notoriously lacks, could apply it to the confusion that normally surrounds moments of crisis. She is organized, clear-headed, and able to prioritize instantly and well. But she is untested. She has never had the ultimate authority for running any organization, for responding to any serious crisis.

And there is another surprisingly vulnerable side of Hillary that should be considered. Although it is rarely shown in public and usually held in check, she often succumbs to tears in times of adversity. Surprisingly, this tough, combative, sharp-tongued woman often broke down in tears during her time as first lady. Quite often, her tears were mixed with anger, as they were at various points with me, George Stephanopoulos, and Rahm Emanuel. At other times they were tears of frustration and sadness, in moments when she felt defeated or incompetent. I witnessed this several times when Hillary seemed bewildered by the turn of events in the health care debate. The stereotype of an emotionally volatile woman would obviously work against her. On the other hand, President Bush has been shown shedding tears at highly emotional moments, so perhaps it is now becoming acceptable for the commander-in-chief to show his or her feelings. In times of crisis, however, it could certainly be a problem.

In sharp contrast to Hillary, Rice has been dealing with crisis for decades in the world of foreign policy and would come to office with more experience at dealing with explosive global situations than any president since Eisenhower.

And finally, how would each grow in office?

Rice demonstrated a tremendous capacity for growth in how well she rose to the occasion at Stanford on becoming its provost and in the way she adjusted to the new environment of 9/11. The cold warrior/diplomat of 1990 became a national leader after that dismal September day. Her evolving understanding of the need for a morally grounded foreign policy that rotates around the push for global democracy shows how much this woman can grow to meet new demands and situations.

Hillary's inability to accept criticism makes it harder for her to grow. She is good at giving the appearance of growth—just as Nixon kept trotting out "new Nixons" at each stage of his career—but there is little evidence that she really does grow. She often comes to tactical conclusions, altering her conduct to suit the opportunities and challenges of the moment—as when, for example, she withdrew from day-to-day management of the White House to a life of traveling and writing—but these shifts are not really growth in the sense of an expanded horizon or evolving character.

She is capable of surprising us. Since her election to the Senate in November 2000, Hillary Clinton seems happier than she did in the White House. She smiles more and seems less emotionally fragile than she did as first lady. At this point it's still hard to know whether these changes are real or just a manifestation of a new version of Hillary. But certainly the experience of being a duly elected senator and no longer being identified as just the wife of a governor or president must be empowering to her. No longer is her power and celebrity derivative of his. Now she finally has her own space, her own life, her own politics, her own position, and for one who has sought it for so long, that must be a heady relief.

During her years as first lady, Hillary had a reputation for being in a near-constant state of rage. She was sullen and bitter. Everyone was her enemy—the press, the Right wing, insurance companies, the women, Ken Starr, the Whitewater prosecutors. As the *Washington Post's* John Harris writes in his recent book, *The Survivor,* "Hillary Clinton was often crabby when accompanying her husband abroad, or so it seemed to aides. She loved travel but did not like the second-fiddle role that was inevitably assigned to her when she traveled with him. On presidential travel . . . she was at the mercy of his schedule, and was subjected to all manner of teas, receptions, and other ceremonial functions."[2] Harris also describes a speech given by Hillary in Australia in which she used "acrid humor"[3] to illustrate her frustration at the criticisms of her activist role. Not exactly the subject or tone one would expect for a first lady speaking in a foreign land.

Starting with her campaign for Senate, though, Hillary seems to have lightened up. Having moved into her own political career, it's possible that she may finally have absorbed the lessons it has to teach. She may have seen the limits of confrontational politics and the virtual certainty that anything that is hidden in Washington will always come out—and come back to haunt you.

Then again, she may not. Beneath the newfound good cheer, she may be the same old Hillary she has always been. We just don't know.

Each candidate, of course, would be a first—the first woman to hold the presidency. Rice would be a double first—the first African American as well.

But the effect of Condi's candidacy and her success would be far more important than her presidency itself could possibly be. Her election would eliminate the last lingering division that threatens our social cohesion.

The election of 2008 will be the next great presidential race. With the possibility of two popular women as candidates, the voters will make history.

We can only hope it's the right kind of history.

NOTES

Chapter 1: Setting the Stage

1. 52 percent now support her candidacy: Frank Newport, "Update: Hillary Clinton and the 2008 Election," June 7, 2004, www.gallup.com/poll/content/login.aspx?ci=16651.
2. "would be likely": Susan Page, "Poll Majority Say They'd Be Likely to Vote for Clinton," *USA Today,* May 26, 2005, www.usatoday.com/news/washington/2005-05-26-hillary-poll_x.htm.
3. a woman for president in 2008: Ibid.
4. And the voters like them both: Ibid.
5. the "Hillary Brand": Dick Morris with Eileen McGann, *Rewriting History* (New York: ReganBooks, 2004), pp. 33–70.
6. Their potential battle recalls: "Action Between USS *Monitor* and CSS *Virginia,* 9 March 1862," *Naval Historical Center,* www.history.navy.mil/photos/events/civilwar/n-at-cst/hr-james/9mar62.htm.
7. seventy thousand dollars in: Barbara Olsen, *The Final Days: The Last Desperate Abuse of Power by the Clinton White House* (Washington, D.C.: Regnery, 2001), pp. 68–69.
8. "I have never wanted": Bill Sammon, "2008 Run, Abortion Engage Her Politically," *Washington Times,* March 12, 2005, www.washtimes.com/national/20050311-115948-2015r.htm.
9. "If nominated": "William Tecumseh Sherman," *PBS.org,* www.pbs.org/weta/thewest/people/s_z/sherman.htm.
10. "Secretary of State Condoleezza Rice": Steven R. Weisman, "Rice on a Run for President: No, Nyet, Nein," *New York Times,* March 14, 2005,

www.nytimes.com/2005/03/14/politics/14rice.html?ex=1123041600&en=5fe58659d022556e&ei=5070&oref=login.

11. "War is the continuation . . ." Carl von Clausewitz, *On War* (New York: Viking Press, 1982), p. 464.

12. The process seems already: Robert D. Blackwell, "Diplomacy Is Back at the State Department," *Wall Street Journal,* July 22, 2005, www.freerepublic.com/focus/f-news/1448243/posts.

Chapter 2: President Hillary: How It Could Happen

1. 33 percent of conservatives: Susan Page, "Poll Majority Say They'd Be Likely to Vote for Clinton," *USA Today,* www.usatoday.com/news/washington/2005-05-26-hillary-poll_x.htm.

2. In her Senate race in 2000: Raymond Hernandez, "Senator Clinton Piles Up a Fund-Raising Lead for 2006," *New York Times,* April 19, 2005, p. A21, http://stophillarypac.com/cgi-data/news/files/51.shtml.

3. "Democratic Party's single best": Adam Nagourney and Raymond Hernandez, "For Hillary Clinton, a Duel Role as Star and as Subordinate," *New York Times,* late ed., October 22, 2002, p. A1.

4. raised at least 45 million: Hernandez, "Senator Clinton Piles Up a Fund Raising Lead for 2006," *New York Times,* April 19, 2005, p. A21, http://stophillarypac.com/cgi data/news/files/51.shtml.

5. Now that Hillary is gearing up: Ibid.

6. $7.5 million to Americans Coming Together: "America Coming Together—Nonfederal Account," *The Center for Public Integrity,* www.public-i.org/527/search.aspx?act=com&orgid=649.

7. sold at least 1,324,727 copies: http://book.bookscan.com/win-cgi/book.exe?what=26595.

8. even the $200,000: "What is the President's Salary?" www.govspot.com, www.govspot.com/know/presidentsalary.htm.

9. erase from her biography: For a full understanding of just how Hillary reinvented herself, see Dick Morris with Eileen McGann, *Rewriting History* (New York: ReganBooks, 2004).

10. rating below 40 percent: Fox News, Opinion Dynamics Survey, March 10, 2001.

11. a 53 percent favorability: Ibid., June 14–15, 2005.

12. Hillary now runs far ahead: Ibid.

13. "Senator Clinton," he told: Katherine Q. Seelye, "Gingrich Sees Mrs. Clinton as Presidential Nominee in 2008," *New York Times,* late ed., April 14, 2005, p. A22.

14. 38 percent of voters: Fox News, Opinion Dynamics Survey, June 14–15, 2005.

15. "the smartest American politician": Seelye, "Gingrich Sees Mrs. Clinton," p. A22.

16. Clinton warned Kerry: Evan Thomas, "The Inside Story: How Bush Did It," *Newsweek, Special Election Issue: Campaign 2004,* November 15, 2004, p. 102, www.msnbc.msn.com/id/3144249/site/newsweek.

17. Look at the numbers: "2004 U.S. Presidential Election Results," CNN.com, www.cnn.com/ELECTION/2004/pages/results/president.

18. In the 2004 race: "2004 U.S. Presidential Election: National Exit Poll," CNN.com, www.cnn.com/ELECTION/2004/pages/results/states/US/P/00/epolls.0.html.

19. 10 percent of the vote: "2000 U.S. Presidential Election: National Exit Poll," CNN.com, www.cnn.com/ELECTION/2000/results/index.epolls.html.

20. jumped to 12 percent: "2004 U.S. Presidential Election: National Exit Poll," CNN.com, www.cnn.com/ELECTION/2004/pages/results/states/US/P/00/epolls.0.html.

21. Hispanics voted for Gore: "2000 U.S. Presidential Election: National Exit Poll," CNN.com, www.cnn.com/ELECTION/2000/results/index.epolls.html.

22. Hillary would likely bring back: "Hillary's New York State of Mind," CBS News, November 8, 2000, www.cbsnews.com/stories/2000/11/03/politics/main246677.shtml.

23. American Hispanics cast only six percent: "2000 U.S. Presidential Election: National Exit Poll," CNN.com.

24. risen to 8 percent: "2004 U.S. Presidential Election: National Exit Poll," CNN.com.

25. Hispanics now account: "Population of the United States by Race and Hispanic/Latino Origin, Census 2000 and July 1, 2004," www.infoplease.com, www.infoplease.com/ipa/A0762156.html.

26. Bush defeated Gore among white women: "2000 U.S. Presidential Election: National Exit Poll," CNN.com.

27. But in 2004 he destroyed Kerry: Ibid.

28. 8.4 million more women voted: Ibid.

29. Women are more likely: "The Gender Gap and the 2004 Women's Vote Setting the Record Straight," Rutgers University, Center for American Women and Politics, www.cawp.rutgers.edu/Facts/Elections/GenderGapAdvisory04.pdf.

30. In 2004, married women backed: Ibid.

31. 42 percent of all registered female voters: Ibid.

32. In New York State, where she has been: Statewide survey of 1,000 adults, Fox News/Mason-Dixon New York State Poll, May 14–15, 2005.

33. almost all of the Kerry vote: Fox News, Opinion Dynamics Poll, June 14–15, 2005.

34. Bush beat Kerry by only: "2004 U.S. Presidential Election: National Exit Poll," CNN.com.

35. The *USA Today*/CNN/Gallup Poll found that: Page, "Poll Majority Say." *USA Today.*

36. sixty percent of New Yorkers: "Clinton Buries All Challengers 2–1 or More, Quinnipiac University Poll Finds; Most New Yorkers Want Her to Serve Full 2nd Term," Quinnipiac University Polling, www.quinnipiac.edu/x11373.xml?ReleaseID=680.

37. "I am focused on winning re-election": Carl Limbacher, "Hillary Declines to Take the Pledge," *NewsMax.com,* May 27, 2005, www.newsmax.com/archives/ic/2005/5/27/132845.shtml.

38. Another recent poll, this time by Marist College: Dr. Lee M. Miringoff, "New York State: Outlook Bright for Senator Clinton in '06 . . . Voters Not Sold on NYS Pols' '08 Prospects," Marist College Institute for Public Opinion, April 12, 2005, www.maristpoll.marist.edu/nyspolls/HC050412.htm.

39. as governor of Texas: Raymond Hernandez, "One Clinton, at Least, Finds the Race in 2008 Worth Discussing," *New York Times,* late ed., June 3, 2005, p. B2.

40. "Why would Hillary": Author interview with former DNC Chairman Terry McAuliffe, September 1, 2004.

Chapter 3: How Condi Can Beat Hillary

1. "They are two brilliant women,": Author interview with former congressman Mike Espy, April 23, 2005.

2. "very strong candidate": Author interview with editor Jamal E. Watson, *Amsterdam News,* May 31, 2005.

3. "A ten-point swing": Author interview with Sean Hannity, Fox News, May 31, 2005.

4. until Eisenhower in 1952: Michael Zak, *Back to Basics for the Republican Party* (Maryland: Signature Books, 2003), p. 163.

5. "My country 'tis of thee": Ibid., pp. 171–172.

6. Jackie Robinson, was a Republican: Ibid., pp. 189–190.

7. "My father joined our party": "Text: Condoleezza Rice at the Republican National Convention," WashingtonPost.com, August 1, 2000, www.washingtonpost.com/wp-srv/onpolitics/elections/ricetext080100.htm.

8. On election day, blacks showed their: Zak, *Back to Basics,* p. 195.

9. In the House, 80 percent: Ibid., p. 202.

10. In opposing this landmark legislation: Ibid.

11. "mildly pro-choice": Transcript of Secretary of State Condoleezza Rice's interview with editors and reporters at the *Washington Times,* www.washingtontimes.com/world/20050311-102521-9024r.htm.

Chapter 4: Being Condoleezza Rice

1. "Be not afraid of greatness": William Shakespeare, *Twelfth Night,* Act II, Scene V, www.william-shakespeare.info/act2-script-text-twelfth-night.htm.
2. "several of the crucial": Nicholas Lemann, "Without a Doubt: Has Condoleezza Rice Changed George W. Bush, or Has He Changed Her?" *New Yorker,* October 14, 2002, www.newyorker.com/printables/fact/021014fa_fact3.
3. "I hope each student": Condoleezza Rice, "Campus Viewpoint: Doing What's Important Before We All Leave Stanford," *Stanford Daily,* May 20, 1999.
4. the highest honor granted to the: Antonia Felix, *Condi: The Condoleezza Rice Story* (New York: Newmarket Press, 2005), p. 84.
5. "a tea for the new teachers": Ibid.
6. at Constitution Hall, in Washington: Lemann, "Without a Doubt."
7. In June, 2005 she played: "Secretary Rice Takes Concert Spotlight," *VOA News,* June 12, 2005, www.voanews.com/English/2005-06-12-voa23.cfm.
8. "to fall in love": Lemann, "Without a Doubt."
9. "He was nothing but": Felix, *Condi,* p. 97.
10. "a few months after she arrived": Ibid., p. 115.
11. "this is somebody I need": Lemann, "Without a Doubt."
12. "greatly impressed by her": "Casper selects Condoleezza Rice to be next Stanford provost," *Stanford University News Service,* news release, May 19, 1993, www.stanford.edu/dept/news/pr/93/930519Arc3267.html.
13. "The governor and Condi hit if off immediately," Felix, *Condi,* pp. 8–9.

14. "One day, I'll be in that house": Lemann, "Without a Doubt."
15. not "college material": Laura B. Randolph, "Black women in the White House: three trailblazers make history at the highest level of American power," *Ebony,* October 1990, www.findarticles.com/p/articles/mi_m1077/is_n12_v45/ai_8904380.
16. on Rice's "quiet demeanor": Barbara Slavin, "Rice Called a Good Fit for Foreign Policy Post," *USA Today,* December 18, 2000, p. 2A, www.usatoday.com/news/vote2000/bush42.htm.
17. "The roadside is littered": Romesh Ratnesar, "Condi Rice Can't Lose," *Time,* September 20, 1999, http://edition.cnn.com/ALLPOLITICS/time/1999/09/20/rice.html.
18. "With a smile on her face": Ibid.
19. "a certain condescension": Mark Z. Barabak, "Condoleezza Rice at Stanford," *Los Angeles Times,* January 16, 2005, http://hnn.us/roundup/comments/9732.html.
20. "gave the impression of": Lemann, "Without a Doubt."
21. When Condi's father went to register: Felix, *Condi,* p. 57.
22. "I remember"": Ibid., p. 45.
23. "These terrible events burned": Ibid., p. 56.
24. "when a firebomb landed": Ibid., p. 50.
25. "The people there stopped": Ibid., p. 58.
26. "Rather than crouch down": Ibid., p. 70.
27. "Let's get one thing": Lemann, "Without a Doubt."
28. "She was an only child": Ibid.
29. "brought a special intensity": Ibid.
30. "Our parents really did": "Condoleezza Rice," biography, galeschools.com, www.galeschools.com/black_history/bio/rice_c.htm.
31. "blast through the barriers": Shalini Bhargava, "Provost Going Back to Her Passions," *Stanford Daily,* January 4, 1999, www.stanforddaily.com/tempo?page=content&id=4220&repository=0001_article.

32. "how her paternal grandfather": "Condoleezza Rice: Profile of a 'Velvet Hammer': From Alabama, to Stanford, Then Maine and White House," Associated Press, April 7, 2004, www.msnbc.msn.com/id/4684024.
33. "very accomplished professional people": Lemann, "Without a Doubt."
34. "My father was not": Felix, *Condi,* p. 55.
35. "To say that John Rice was": Ibid., pp. 46–47.
36. he got his masters degree: Ibid., p. 60.
37. "my parents were very": Ibid., p. 43.
38. "I was lucky to have": Hillary Rodham Clinton, *Living History* (New York: Simon & Schuster, 2003), p. 20.
39. "in a cautious, conformist era": Ibid., p. 14.
40. the television show "Happy Days": Ibid., p. 18.
41. watch "Ed Sullivan" on Sundays: Ibid.
42. riding her bike "everywhere": Ibid., p. 13.
43. "allowed to go to the Pickwick Theater": Ibid., p. 14.
44. "Find your passion,": Lemann, "Without a Doubt."
45. "I was elected co-captain": Ibid., p. 15.
46. "asked me to be on": Ibid., p. 19.
47. "ran for student government": Ibid., p. 24.
48. "elected president of the local": Ibid., p. 18.
49. "elected president of our college's": Ibid., p. 31.
50. college student government: Ibid., p. 38.
51. "the great intellectual divide": Lemann, "Without a Doubt."

Chapter 5: Hillary's Senate Record: The Grand Deception

1. "solid freshman term": David Remnick, "Political Porn," *New Yorker,* July 4, 2005, www.newyorker.com/talk/content/articles/050704ta_talk_remnick.

2. Symbolic . . . Substantive Congressional Record for 107th and 108th Congress.

3. "drum up support": Geoff Earle, "Clinton's GOP allies," *Hill,* June 21, 2005, www.hillnews.com/thehill/export/TheHill/News/Frontpage/062105/clinton.html.

4. She teamed up with: Ibid.

5. "When she was first lady,": author interview with Senator Mike Devine, May 27, 2005.

6. According to the semi-official tally in *Congressional Quarterly,:* "Hillary's Senate record," *Washington Times,* editorial, op-ed, November 21, 2004, www.washtimes.com/op-ed/20041120-084025-3316r.htm.

7. "I grew so rich that I was sent": *HMS Pinafore,* Gilbert and Sullivan.

8. The AFL-CIO says: Senator voting record on Labor issues, "Senator Hillary Rodham Clinton (NY)," *Project Vote Smart,* www.votemart.org/issue_rating_category.php?can_id=WNY99268&type=category&category=Labor.

9. The Americans for Democratic Action: Compiled by Congressional Quarterly based on information provided by each participating group. Find interest group descriptions at http://cq.com/display.do?dockey=/cqonline/prod/data/docs/html/member/member-0000007201.html@member&metapub=CQMEMBER&searchIndex=1&seqNum=1.

10. While all Democrats put together: Greg Sargent, "Brand Hillary," *Nation,* June 6, 2005, www.thenation.com/doc.mhtml?i=20050606&c=5&s=sargent.

11. Hillary Clinton's votes all echo the liberal line: Congressional Record for 107th and 108th Congress.

12. "The bill was going": Author interview with Senate staff member, April 4, 2005.

13. "supported the 'No Child Left Behind'": Education section of Senator Hillary Rodham Clinton's official website, http://clinton.senate.gov/issues/education/index.cfm?topic=elementary.

14. "topped the Senate": Annie Patnaude, Demian Brady, and Peter J. Sepp, "Study: Nearly 10 Years After GOP Takeover, Trend to Spend Still Shaping Congressional Agendas," NTUF Press Releases, October, 7, 2004, www.ntu.org/main/press_release_printable .php?PressID=653&org_name=NTUF.

15. "facilitated" a $93 million: Accomplishment section of Senator Hillary Rodham Clinton's official website, http://clinton.senate.gov /accomplishments_108th.html.

    (This information, and that in other citations marked with an asterisk below, was drawn from pages on Hillary Rodham Clinton's official website dealing with her Senate record. However, changes to her website have since rendered the URLs given here obsolete. We have preserved a complete text of the quoted material as it originally appeared on her website.)

16. the Broadband conference in Delhi: http://clinton.senate.gov /accomplishments_108th.html.*

17. She boasts of co-sponsoring: http://clinton.senate.gov /accomplishments_108th.html.*

18. She gives herself credit: http://clinton.senate.gov/accomplishments _108th.html.*

19. She says she "visited Israel": http://clinton.senate.gov /accomplishments_108th.html.*

20. "Ford to City": Wikipedia, "New York *Daily News*," www.answers .com/topic/new-york-daily-news.

21. "At that moment": Hillary Clinton interview with Jane Pauley, *Today*, September 18, 2001, http://perec.wnyc.org/blog/lehrer /archives/000064.html.

22. "Schumer spent Wednesday morning deploying his staff,": Carl Limbacher, "Brill to NewsMax: Notes Prove Hillary Made Up 9/11 Role," NewsMax.com, April 28, 2003, www.newsmax.com/archives /articles/2003/4/27/213024.shtml.

23. "I hear you want to": Carl Limbacher, "Steven Brill: Hillary Fabricated 9/11 Records," NewsMax.com, April 20, 2003, http:// newsmax.com/scripts/showinside.pl?a=2003/4/20/143224.

24. "the person responsible for": Ibid.

25. "an elaborate story": Ibid.

26. "None of it turned": Ibid.

27. "it sort of takes": Ibid.

28. "I think [Sen. Clinton]": Ibid.

29. "This family had tried repeatedly": Ibid.

30. "Meanwhile," said Brill, "Senator Schumer":Ibid.

31. "What stunned me": Ibid.

32. "Brill's accusations are": Carl Limbacher, "Brill to Hill: Release 9/11 Meeting Logs," NewsMax.com, May 6, 2003, http://newsmax.com/scripts/showinside.pl?a=2003/5/6/93726.

33. "If Hillary Clinton": Ibid. Carl Limbacher, "Brill to NewsMax: Notes Prove Hillary Made Up 9/11 Role," NewsMax.com, April 28, 2003, www.newsmax.com/archives/articles/2003/4/27/213024.shtml.

Chapter 6: Rice at the Pinnacle

1. "I had chosen Condi": John Prados, "Blindsided or blind," July/August 2004, pp. 27–37, www.thebulletin.org/article.php?art_ofn=ja04prados.

2. The dichotomy between the two approaches: Henry Kissinger, *Diplomacy* (New York: Simon and Schuster, 2004), chapters 1 and 2.

3. "came to see": Antonia Felix, *Condi: The Condoleezza Rice Story* (New York: Newmarket Press, 2005), p. 95.

4. "I hope you know a lot": Romesh Ratnesar, "Condi Rice Can't Lose," *Time,* September 20, 1999, http://edition.cnn.com/ALLPOLITICS/time/1999/09/20/rice.html.

5. "The most frightening moments": "Interview with Dr. Condoleezza Rice," National Security Archive, George Washington University, December 17, 1997.

6. "Gorbachev said, 'We want the United States' ": Ibid.

7. "Events were unfolding": Ibid.

8. "When you have": Jacob Heilbrunn, "The Unrealistic Realism of Bush's Foreign Policy Tutors–Team W," *The New Republic,* September 27, 1999.
9. "Academic friends who": Nicholas Lemann, "Without a Doubt: Has Condoleezza Rice Changed George W. Bush, or Has He Changed her?" *The New Yorker,* October 14, 2002, www.newyorker.com/printables/fact/021014fa_fact3.
10. "to miss . . . the revocation": Prados, "Blindsided or Blind."
11. Rice helped convince Bush: Felix, *Condi,* p. 143.
12. "the Bush Administration seemed": Heilbrunn, "The Unrealistic Realism."
13. "Germany unification was": "Interview with Dr. Condoleezza Rice," National Security Archive.
14. "Rice saw the problem": Prados, "Blindsided or Blind."
15. "in a race to try to"": "Interview with Dr. Condoleezza Rice," National Security Archive.
16. "be respectful of Soviet interests": Ibid.
17. "Was it inevitable": Felix, *Condi,* pp. 149–150.
18. "We will maintain": Lemann, "Without a Doubt."
19. "the foreign policy staff": Laura Flanders, "Never Apologize," adapted from her book, *Bushwomen: Tales of a Cynical Species* (Verso, March 2004).
20. "He struck me as mercurial": Lemann, "Without a Doubt."
21. "This isn't the door": Felix, *Condi,* pp. 144–145.
22. "As an executive": Ibid., pp. 13–14.
23. "The school acknowledged": Mark Z. Barabak, "Condoleezza Rice at Stanford," *Los Angeles Times,* January 16, 2005, http://hnn.us/roundup/comments/9732.html.
24. "Stanford had a history": Ibid.
25. "It would be disingenuous": Lemann, "Without a Doubt."
26. "her age was clearly": Bruce Anderson, "An Interview with Gerhard Casper," *Stanford Magazine,* September 1993.

27. more of "deputy president": Ibid.

28. "the crucial person": Ibid.

29. "required grit, skill, political,": Anderson, "An Interview with Gerhard Casper."

30. "a candidate for provost": Felix, *Condi,* p. 172.

31. "the toughest": Lemann, "Without a Doubt."

32. $20 million annual budget deficit: Felix, *Condi,* p. 173.

33. cut almost $40 million: Ed Guzman and Adam Kemezis, "Budget Cuts, Student Relations Highlight Six Outstanding Years as Provost," *Stanford Daily,* January 4, 1999.

34. "There was sort of conventional": Felix, *Condi,* p. 174.

35. "No, we're going to": Ibid.

36. "I actually don't think": Ibid.

37. "No one likes layoffs": Barabak, "Condoleezza Rice at Stanford."

38. "I don't do committees": Ibid.

39. "Maybe I was too": Lemann, "Without a Doubt."

40. "She's very charming": Ibid.

41. "I'm not hungry": Ibid.

42. "As the university's No. 2": Barabak, "Condoleezza Rice at Stanford."

43. Indeed, Rice's administration was so successful: Felix, *Condi,* p. 180.

44. A 1998 Stanford report by the Faculty: Paula Findlen, Estelle Freedman, Nancy Kollmann, Cecilia Ridgeway, Mary Louise Roberts, Debra Satz, "The Status of Women on the Stanford Faculty Report to the Faculty Senate," *The Faculty Women's Caucus,* Spring 1998.

45. passed Proposition 209: "Analyzing the Impact of Proposition 209 in California Higher Education," www.landmarkcases.org.

46. "I am myself": Diane Manual, "Senators, Others Debate Status of Women Faculty," *Stanford Report,* May 20, 1998, http://newsservice.stanford.edu/news/1998/may20/facsen520.html.

47. "I support affirmative action": Shalini Bhargava, "Provost Going Back to Her Passions," *Stanford Daily,* January 4, 1999, www.stanforddaily.com/tempo?page=content&id=4220&repository=0001_article#.

48. "an extremely broad": Diane Manuel, "Hiring, Tenuring of Women Faculty Topic of Two Reports to Faculty Senate," *Stanford Report,* May 13, 1998, http://news-service.stanford.edu/news/1998/may13/womfacsen513.html.

49. "a real slippery slope": Guzman and Kemezis, "Budget Cuts, Student Relations."

50. "We have a three-year": Felix, *Condi,* p. 116.

51. Akhil Gupta: Audrey Harris, "Tenure for Women Continues to Cause Concern," *Stanford Daily,* July 25, 2002, www.stanforddaily.com/tempo?page=content&id=8657&repository=0001_article.

52. Karen Sawislak: Ibid.

53. Linda Mabry: Bill Workman, "Shadow on Diversity at Stanford: Spate of Faculty Departures Dismays Black Law Students," *San Francisco Chronicle,* February 15, 1999, www.sfgate.com/cgi-bin/article.cgi?file=/chronicle/archive/1999/02/15/MN55114.DTL.

54. Robert Warrior: "Robert Warrior denied tenure in English Department at Deans' Level," *Stanford Report,* February 24, 1999, www.stanford.edu/group/aware/diversity/warrior_denied_tenure.htm.

55. forty-six new minority: Kimberly Downs, "Senate Discusses Female Faculty," *Stanford Daily,* May 14, 1999, http://daily.stanford.edu/tempo?page=content&id=2853&repository=0001_article.

56. By the time Rice left her post: Findlen et al., "The Status of Women."

57. By the end of Rice's term as provost women were getting tenured: "Senate Discusses Female Faculty," *Stanford Daily,* May 14, 1999, http://daily.stanford.edu/tempo?page=content&id=2853&repository=0001_article.

58. had a women dean: Ibid.

59. "widespread discontent with": Dana Mulhauser, "Hennessy Takes Flight: Future Provost Will Bring His Problem-Solving Skills to Bear," *Stanford Daily,* May 26, 1999, http://daily.stanford.edu/tempo?page=content&id=2721&repository=0001_article.

60. "Improbably, the youngest provost": Barabak, "Condoleezza Rice at Stanford."

61. "it appears the Stanford": Fred Luminoso and Louise Auerhahn, Students for Environmental Action at Stanford, "A Culture of Bias," http://seas.stanford.edu/diso/articles/cultureofbias.html.

62. When she left the university: "Farewell, Provost Rice: Condi Leaves a Legacy as a Powerful Administrator Who Cares About Students," editorial, *Stanford Daily,* January 5, 1999, http://daily.stanford.edu/daily/servlet/Story?id=4306§ion=News&date=01-05-1999.

63. "We've done more": Mulhauser, "Hennessy Takes Flight."

64. "I don't suffer from": Felix, *Condi,* p. 187.

65. "Superficially, Bush and Rice": Evan Thomas, "US Politics: The Quiet Power of Condi Rice: Born in 'Bombingham,' the Enigmatic Adviser Has Become the 'Warrior Princess'—Bush's Secret White House Weapon," *Newsweek,* December 11, 2002, http://bulletin.ninemsn.com.au/bulletin/EdDesk.nsf/0/c13b21d1f20c61a7ca256c89007556dc?OpenDocument.

66. "Well, first, my faith": Bill Sammon, "2008 Run, Abortion Engage Her Politically," *Washington Times,* March 12, 2005, www.washtimes.com/national/20050311-115948-2015r.htm.

67. "During the campaign": Thomas, "US Politics."

68. "Rice not only works": Lemann, "Without a Doubt."

69. "Rice is quiet": Thomas, "US Politics."

70. "her job as national-security advisor": Ibid.

71. "Rather than her simply": Lemann, "Without a Doubt."

72. "Bush's moral impulses": Thomas, "US Politics."

73. "as power diffuses": Heilbrunn, "The Unrealistic Realism."

74. "Power matters": Jay Nordlinger, "Star-in-Waiting: Meet George W's Foreign-Policy Czarina," *National Review,* August 30, 1999, www.findarticles.com/p/articles/mi_m1282/is_16_51/ai_55432936.

75. "A balance of power that supports freedom,": Condoleezza Rice, remarks to the International Institute for Strategic Studies, June 26, 2003, www.whitehouse.gov/news/releases/2003/06/20030626.html.

76. "Yes, people want": Condoleezza Rice, commencement address at Stanford, *Stanford Report,* June 16, 2002, http://news-service.stanford.edu/news/2002/june19/comm_ricetext-619.html.

77. "Realists play down": Condoleezza Rice, "America Has the Muscle, but It Has Benevolent Values, Too," *London Telegraph,* October 17, 2002, www.freerepublic.com/focus/f-news/770486/posts.

78. "received more than forty": Transcript of Rice's 9/11 Commission statement, May 19, 2004, www.cnn.com/2004/ALLPOLITICS/04/08/rice.transcript.

79. "did not raise the possibility that terrorists might": Ibid.

80. In our book *Because He Could:* Dick Morris and Eileen McGann, *Because He Could,* ReganBooks, New York, 2004.

81. "Rice found that her": B. Denise Hawkins, "Condoleezza Rice's Secret Weapon: How our National Security Adviser finds the strength to defend the free world," *Today's Christian,* September/October 2002, Vol. 40, No. 5, p. 18, www.christianitytoday.com/tc/2002/005/1.18.html.

82. "Prayer is important": Sammon, "2008 Run."

83. "I try always to": Condoleezza Rice, "Walk of Faith," speaking to a Sunday school class at National Presbyterian Church, August 4, 2002, printed in the *Washington Times,* August 27, 2002, http://chebar0.tripod.com/id117.htm.

84. Rice offered a window into her approach: Sheryl Henderson Blunt, "The Privilege of Struggle—How Rice understands suffering and prayer," *Christianity Today,* Vol. 47, No. 9, p. 44, September 2003, www.christianitytoday.com/ct/2003/009/33.44.html.

85. "certain moral clarity": Hawkins, "Condoleezza Rice's Secret Weapon."

86. "I feel that faith": Rice, "Walk of Faith."

87. "I've watched over": Ibid.

88. "generosity of spirit": Hillary Rodham Clinton, *Living History* (New York: Simon & Schuster, 2003), pp. 135–136.

89. "I've always been": Michael Jonas, "Sen. Clinton Urges Use of Faith-based Initiatives," *Boston Globe,* January 20, 2005, www.boston.com/news/local/massachusetts/articles/2005/01/20/sen_clinton_urges_use_of_faith_based_initiatives?mode=PF.

90. "wondering if it was": John Harris, "Applying the Salve of Prayer; Clintons Use Gathering to Speak Out Against Anger, Cynicism," *Washington Post,* February 7, 1997, p. A01.

91. "I didn't hit my stride": Clinton, *Living History,* pp. 27–28.

Chapter 7: The Two Hillarys: Dr. Jekyll and Mrs. Hyde

1. "If he vetoed welfare reform a third time,"": Hillary Rodham Clinton, *Living History* (New York: Simon & Schuster, 2003), p. 369.

2. The *National Journal* evaluated Hillary's voting record: Richard E. Cohen, "How They Measured Up," *National Journal,* February 28, 2004, www.johnsullivanforcongress.com/news_articles/2003_vote_rankings.pdf.

3. "You can fool": Abraham Lincoln, *Quote DB,* www.quotedb.com/quotes/4183.

4. "view the former": Scott Rasmussen, "72% Say They're Willing to Vote for Woman President," survey of 1,000 adults, *Rasmussen Reports,* April 6–7, 2005, www.rasmussenreports.com/2005/Woman%20President.htm.

5. "Gingrich says he": Raymond Hernandez, "New Odd Couple: Hillary Clinton and Newt Gingrich," *New York Times,* May 13, 2005, p. 1, www.iht.com/articles/2005/05/13/news/clinton.php.

6. "I know it's a bit": Ibid.

7. She criticizes Gingrich for falsely claiming: Clinton, *Living History,* p. 232.

8. she blames him for protesters who appeared: Ibid., p. 246.

9. for his "glee": Ibid., p. 257.

10. children of unwed mothers: Ibid., p. 262.

11. Hillary as a "bitch": Ibid., p. 263.

12. "pontificat[ed] about American history": Ibid.

13. "You know he": Ibid.

14. "Perhaps it was someone's": Ibid., p. 304.

15. his own marital infidelity: Ibid., p. 450.

16. "Hillary Rodham Clinton is": Charles Hurt, "Hillary Goes Conservative on Immigration," *Washington Times,* December 13, 2004, http://washingtontimes.com/national/20041213-124920-6151r.htm.

17. "think that we have": Ibid.

18. "I am, you know,": Ibid.

19. "seriously flawed": "Carl Limbacher, Hillary 'Outraged' Over Real ID Act," NewsMax.com, May 12, 2005, www.newsmax.com/archives/ic/2005/5/12/124534.shtml.

20. "moral issues": "Voters Liked Campaign 2004, but Too Much 'Mud-Slinging' Moral Values: How Important?" Pew Research Center, November 11, 2004, http://people-press.org/reports/display.php3?ReportID=233.

21. "If there was doubt": Elaine Monaghan, "Hillary in Starting Blocks for White House Run," *The Times,* January 27, 2004,www.timesonline.co.uk/article/0,,11069-1458269,00.html.

22. "I am and always": Ibid.

23. "respect those who": Ibid.

24. "faith based" initiatives: Ibid.

25. "a silent epidemic": Carl Limbacher, "Hillary: Too Much Sex on TV," NewsMax.com, March 10, 2005, www.newsmax.com/archives/ic/2005/3/10/103917.shtml.

26. In those early days, her polls plunged: Fox News/Opinion Dynamics Survey, January 10–11, 2001 and March 14–15, 2001.

27. "For eight years": Raymond Hernandez, "Not the Mrs. Clinton Washington Thought It Knew," *New York Times,* January 24, 2002, http://query.nytimes.com/gst/abstract.html?res=F20A12F8355F0C778EDDA80894DA404482.

28. She disarmed her Senate colleagues with her: Ibid.

29. "She didn't come in": Ibid.

30. "the Hillary Clinton": Ibid.

31. "I think she is benefiting": Ibid.

32. "to cultivate a bipartisan": Raymond Hernandez, "As Clinton Wins G.O.P. Friends, Her Challengers' Task Toughens," *New York Times,* March 6, 2004, http://68.166.163.242/cgi-bin/readart.cgi?ArtNum=87015.

33. "Mrs. Clinton had been": Ibid.

34. "There has never been an administration": Patrick D. Healy, "Senator Clinton Assails Bush and GOP at Campaign Fund-Raiser," *New York Times,* June 6, 2005, B1.

35. "Some [Republicans] honestly believe they are": Ibid.

36. "Many of you": Carl Limbacher, "Hillary: 'We're Going to Take Things Away From You,'" NewsMax.com, June, 30, 2004, www.newsmax.com/archives/ic/2004/6/30/91013.shtml.

37. "The president knew what?": "Traces of Terrorism: Excerpts from Senator Clinton's Speech," *New York Times,* May 18, 2002, p. A10, http://query.nytimes.com/gst/abstract.html?res=F30B14FC3C5C0C7B8DDDAC0894DA404482&incamp=archive:search.

38. "I am sick and tired": David Hogberg, "Don't Dare Call Them Unpatriotic," *American Spectator,* May 6, 2003, http://64.233.161.104/search?q=cache:FDz8wsIp6KcJ:www.spectator.org/article.asp%3Fart_id%3D2003_5_5_23_55_48+david+hogberg+a+near+melt-down&hl=en.

39. fourteen courageous moderates: "Senators Defuse Filibuster Feud," CBS News, May 24, 2005, www.cbsnews.com/stories/2005/05/24/politics/main697516.shtml.

Chapter 8: How Hillary Came to Be Hillary

1. "stayed at home and baked cookies": Hillary Rodham Clinton, *Living History* (New York: Simon & Schuster, 2003), p. 109.
2. seventy-eight foreign countries: Ibid., p. xiv.
3. "Research shows the presence": Carl Limbacher, "Hillary: Women Less Corrupt," NewsMax.com, March 8, 2005, www.newsmax.com/archives/ic/2005/3/8/64125.shtml.
4. "The Foreign Affairs Committee is basically focused": Author interview with Michael Herson, March 3, 2005.
5. "Outsourcing will continue": Carl Limbacher, "Hillary in India: 'Outsourcing Will Continue' ": NewsMax.com, February 28, 2004, www.newsmax.com/archives/ic/2005/2/28/104755.shtml.
6. Unfortunately, Hillary was also a sponsor: "Concern over Outsourcing Jobs Sense of Senate," Senate Amendment 2311, 108th Congress.
7. expressing "concern" over his: Vincent Morris, "Hillary Stick to Guns in Feud," *New York Post,* February 26, 2005, www.nypost.com/news/nationalnews/41356.htm.

Chapter 9: But . . .

1. including 260 embassies, consulates: "Department of State and International Assistance Programs," Office of Management and Budget, August 7, 2005, www.whitehouse.gov/omb/budget/fy2005/state.html.
2. During her book tour to promote: Author interview, May 2004.
3. "ran a gas station": Joe Mahoney, "Hil's Gandhi Gas Jockey Joke Fuels Row," New York *Daily News,* January 7, 2004, www.nydailynews.com/front/story/152686p-134376c.html.
4. "went to school": Senator Hillary Rodham Clinton, "Addressing the National Security Challenges of Our Time: Fighting Terror and the

Spread of Weapons of Mass Destruction," Brookings Leadership Forum, February 25, 2004, www.brook.edu/dybdocroot/comm/events/20040225.pdf.

5. "the U.S. is trapped": Carl Limbacher, "Hillary Blasts Bush in Arab Press," NewsMax.com, April 27, 2004, www.newsmax.com/archives/ic/2004/4/27/105458.shtml.

6. set aside $200 million: Sheryl Henderson Blunt, "The Unflappable Condi Rice: Why the World's Most Powerful Woman Asks God for Help," *Christianity Today,* August 22, 2003, www.christianitytoday.com/ct/2003/009/1.42.html.

7. "I agree with the president's": "Condoleezza Rice: The Devil's Handmaiden," commentary, *The Black Commentator,* January 23, 2003, www.blackcommentator.com/26/26_commentary.html.

8. "view that we have to": Transcript of Secretary of State Condoleezza Rice's interview with editors and reporters at the *Washington Times,* www.washingtontimes.com/world/20050311-102521-9024r.htm.

9. "say they would be": Scott Rasmussen, "72% Say They're Willing to Vote for Woman President," survey of 1,000 adults, *Rasmussen Reports,* April 6–7, 2005, www.rasmussenreports.com/2005/Woman%20President.htm.

10. Fourteen women currently: "Snapshots of Current Political Leadership," The White House Project, www.thewhitehouseproject.org/know_facts/snapshots_women.html.

11. fifty-nine serve: "Women Serving in the 108th Congress 2003–05," Center For Women and Politics, www.cawp.rutgers.edu/Facts/Officeholders/cong03.html.

12. "I am a very deeply": Antonia Felix, *Condi: The Condoleezza Rice Story* (New York: Newmarket Press, 2005), p. 19.

13. "going to marry him": Ibid., p. 110.

14. "I really don't know what happened": Ibid.

15. "she would date": Ibid., p. 111.

16. "the NFL job": Ibid., p. 17.

17. Condi would have defeated: Jim Puzzanghera, "Rice Fits Bill for Governor, Some Say," *San Jose Mercury News,* June 18, 2003, www.freerepublic.com/focus/f-news/931618/posts.
18. "This [running for governor] is something": Ibid.
19. "It's not on my radar screen": Ibid.

Chapter 10: 2004: The Year Politics Turned Upside Down

1. But in 2004, books from both sides: Bookscan (www.bookscan.com) is a service that tallies consumer book sales in the United States. Since not all vendors report sales to Bookscan, its statistics may understate actual sales.
2. NewsMax's Web page became the sixth most: Author interview with Chris Ruddy, NewsMax.com, April 20, 2005.
3. "the nine members": "A Red Carpet Entrance for the Palme d'or Winner—23/05/2004," *Festival De Cannes,* www.festival-cannes.fr/films/fiche_film.php?langue=6002&id_film=4201423.
4. a hundred and twenty-five: Author interview with film producer David Bossie, April 23, 2005.
5. Soaring above CNN and MSNBC in its ratings: David Wissing, *Hedgehog Report,* www.davidwissing.com/index.php?s=cable+news+ratings&submit=Search.
6. "Neither [Blades or Boyd] had experience": "About the MoveOn Family of Organizations," MoveOn.org, www.moveon.org/about.html.
7. "seven people on staff": Joe Trippi, *The Revolution Will Not Be Televised: Democracy, the Internet, and the Overthrow of Everything* (New York: ReganBooks, 2004), p. 2.
8. "fired up supporters": Ibid.
9. "There are no votes": *Newsweek* staff, "Fits and Starts," *Newsweek, Special Election Issue: Campaign 2004,* November 15, 2004, p. 48, www.msnbc.msn.com/id/3144249/site/newsweek.

10. "had burst from": Ibid., pp. 45–49.

11. no bounce in the polls: *Newsweek* staff, *Newsweek, Special Election Issue: Campaign 2004,* November 15, 2004, p. 86, www.msnbc.msn.com /id/3144249/site/newsweek.

12. "the Swift Boat vets": Ibid., p. 89.

13. accusing U.S. troops: Ibid., p. 90.

14. They pooh-poohed the Swift Boat ads: Ibid., p. 91.

15. The debate persisted in the Kerry camp: Ibid., p. 92.

16. "For a moment": Ibid., p. 96.

17. "We saved their ass": Author interview with Don Hewitt, December 1996.

18. doubted their authenticity: *Newsweek* staff, *Newsweek,* p. 96.

19. "The most important": Ibid.

20. "effective control [of the Kerry campaign] to Clinton's": Ibid., p. 101.

21. After Bill Clinton called Kerry from his hospital: Ibid., p. 102.

22. "A lie can travel": Mark Twain, *The Quotations Page,* www.quotationspage.com/quotes/Mark_Twain.

23. "hold to account": "Full transcript of bin Ladin's speech," Aljazeera.net, November 1, 2004, http://english.aljazeera.net/NR /exeres/79C6AF22-98FB-4A1C-B21F-2BC36E87F61F.htm.

24. 1.6 million volunteers: Speech by Ed Gillespie on *National Review* cruise, November 2004.

25. "a head fake": Ibid.

Chapter 11: Who Else Is There?

1. "we actually won": Perry Bacon, Jr., "The Eternal Optimist: John Kerry Is on the Road Again, Listing Excuses for Losing in 2004 and Looking Like a 2008 Campaigner," *Time,* March 28, 2005, www.time.com/time/archive/preview /0,10987,1039699,00.html.

2. "merely a detour": Ibid.
3. "Some Democrats on Capitol Hill": Ibid.
4. "It's been a long time": Ibid.
5. "well-known Democratic operative": Mike Allen, "'Fired Up' Kerry Returning to Senate: Aides Say He Wants to Act as Counter to Bush, and Possibly Run in 2008," *Washington Post,* November 9, 2004, www.washingtonpost.com/wp-dyn/articles/A35224-2004Nov8.html.
6. "I can't imagine": Ibid.
7. "What makes this guy": *Newsweek* staff, *Newsweek, Special Election Issue: Campaign 2004,* November 15, 2004, p. 78, www.msnbc.msn.com/id/3144249/site/newsweek.
8. "the closer Dean got": Ibid., p. 52.
9. "first gay president": Ibid.
10. "not ready for prime time": Ibid.
11. "Rudy leads the field": Fox News, Opinion Dynamics Survey, June 14–15, 2005.
12. New York's annual homicide: "NYC Remains the Safest Large City in the U.S.," newyorkcityvisit.com, December 24, 2004, www.nycvisit.com/content/index.cfm?pagepkey=1411.

Chapter 12: Secretary of State: What Does the Future Hold for Condi?

1. The United States exports only: Bruce Bartlett, "China Trade Deficit Travail," *Washington Times,* August 31, 2003, www.washtimes.com/commentary/20030831-102452-7124r.htm.
2. 60 percent of our petroleum: e-mail to author from former California EPA Secretary Terry Tamminen, May 24, 2005.
3. "can be sequestered": Ibid.
4. we can get: Ibid.
5. a "hydrogen highway": Author interview with Terry Tamminen January 2005.

Chapter 13: Drafting Condi

1. "As I was going": John Bartlett, *Familiar Quotations* (Boston: Little, Brown and Company, 1955), p. 883.
2. But in New Hampshire, Delaware, and Arizona: Filing data from election laws of the various states.
3. "Ike never put himself": Herbert Brownell with John P. Burke, *Advising Ike: The Memoirs of Herbert Brownell* (Lawrence: University of Kansas Press, 1993), p. 94.
4. "a clear-cut call": Richard Norton Smith, *Thomas E. Dewey and His Times* (New York: Simon & Schuster, 1982), p. 581.
5. "Mr. Conservative": Ibid., p. 304.
6. "one of the vast": Ibid., p. 442.
7. beau ideal of the GOP: Ibid., p. 586.
8. "the only Republican who": Ibid., p. 579.
9. "I had observed": Brownell, *Advising Ike,* p. 97.
10. "The Citizens group": Ibid., p. 107.
11. won by 46,661 votes: Ibid., p. 103.
12. "I keenly realize": Smith, *Thomas E. Dewey,* p. 583.
13. than Taft's 24,000: Brownell *Advising Ike,* p. 106.
14. "certainly it was not": Smith, *Thomas E. Dewey,* p. 582.
15. signed "loyal oaths": Ibid., p. 586.
16. At the end of the first ballot: Ibid., p. 119.

Chapter 14: President Clinton? President Rice? What Kind of President Would Each Make?

1. "The economic royalists hate me": Acceptance Speech to Democratic National Convention by Franklin D. Roosevelt, 1936.
2. "Hillary Clinton was often crabby": John Harris, *The Survivor* (New York: Random House, 2005), p. 321.
3. "acrid humor": Ibid., p. 257.

INDEX